MANAGING FOUNDATION ASSETS

AN ANALYSIS OF FOUNDATION INVESTMENT AND PAYOUT PROCEDURES AND PERFORMANCE

Lester M. Salamon
and Kenneth P. Voytek

A Report to the
Council on Foundations

Ph/Ad

The Foundation Center

Library of Congress Cataloging-in-Publication Data
Salamon, Lester M.
    Managing foundation assets: an analysis of foundation investment
and payout procedures and performance / Lester M. Salamon and
Kenneth P. Voytek.
        p.   cm.
    "A report to the Council on Foundations."
    Includes bibliographical references.
    ISBN 0-87954-283-7
    1. Corporations, Nonprofit—Finance.   I. Voytek, Kenneth P.
II. Council on Foundations.   III. Title.
HG4027.65.S25   1989
361.7'632'0681—dc19

Contents

Preface

This initial report of the Council's Fiscal Policies and Practices Project is a milestone for the foundation field. The first comprehensive study ever done of foundation investment performance and payout, it covers an area of vital importance to the continued growth and vitality of organized philanthropy. At the heart of the concerns that led to this research is the need to understand better the balance that grantmakers must strike between current recipients and the claimants of the future. If foundations are to provide funding in the future, they must be prudent stewards of their resources now.

The findings are encouraging. The majority of foundation assets are well managed and are achieving respectable rates of return. But the research reveals room for improvement, especially among smaller foundations. Armed with the results of this study, the Council will be able to develop approaches to strengthen the investment function among foundations in need of such help.

The findings with regard to the payout rates of foundations are also encouraging. Many foundations were able to pay out more than the required 5 percent of their assets to worthy causes, while preserving the real value of their assets. Foundations clearly benefited from the greater flexibility in investment strategies opened up by the change in the payout rule after 1981.

Assessing the impact of the 1981 change in the pay-out rule was a primary reason the Regional Associations of Grantmakers recommended that the Council undertake this research. In 1980 the Council convinced Congress to change the payout rule from a requirement that foundations pay out all of their income or 5 percent of their assets, whichever was larger, to a payout of a flat 5 percent of assets. This change was needed to arrest the decline in the real value of foundation assets, which were shrinking because the payout rule prohibited reinvesting a foundation's investment returns. The Council and the Regional Associations agreed that a study to document the results of the 1981 change would be necessary to understand its impact, to inform policy makers and to guide the Council's future legislative agenda.

The research was designed by Dr. Elizabeth Boris, vice president for research, with the guidance of the Council's Research Committee, then chaired by Jane Dustan, vice president of the Foundation for Child Development. A distinguished advisory committee, chaired by Dr. William Bradley, then president of the Hazen Foundation, was formed to oversee the project. After a competitive process, Dr. Lester Salamon was commissioned to do the research.

The Council is grateful for the work of all of those involved in producing this report—Dr. Salamon and Kenneth Voytek who conducted the research, the foundations that filled out a complicated survey, the 17 grantmakers that supported the research, the advisory committee that freely gave of its time to review the manuscript and provide insights to the author, members of the Research Committee and foundation officers who read and commented on the manuscript, and Dr. Humphrey Doermann, president of the Bush Foundation, who chaired the Research Committee and served on the advisory committee through much of this process.

It is our hope that the information in this report will lead to a better understanding of the importance of foundation investment policies and procedures both inside and outside of the foundation field. A second phase of the project, updating the investment and pay-out figures, is now in progress.

James A. Joseph

Acknowledgments

Work on this report benefited greatly from the calm prodding, helpful questioning, and judicious oversight provided by Dr. Elizabeth Boris, vice president for research at the Council on Foundations, and from the insight and advice offered by the distinguished advisory committee the Council assembled for this project (see following list of members.) Thanks are also owed to Dr. Peter Williamson of Dartmouth College for technical advice on several crucial aspects of the analysis, to Mr. Peter Bernstein for helpful suggestions on wording and interpretation of the data, to Ms. Margo Brinkley and the staff of the Washington office of the Foundation Center for assistance in developing the financial data on foundations that is the basis for much of our analysis, and to Ms. Jacquelyn Perry for production assistance on the multiple revisions that this manuscript endured. Finally, we are grateful to the numerous foundations that helped support this work, including:

Edna McConnell Clark Foundation
New York, New York

Geraldine R. Dodge Foundation
Morristown, New Jersey

Dyer-Ives Foundation
Grand Rapids, Michigan

Equitable Life Assurance Society of the United States
Equitable Foundation
New York, New York

Ford Foundation
New York, New York

Edward W. Hazen Foundation
New York, New York

William Randolph Hearst Foundation
New York, New York

Conrad N. Hilton Foundation
Los Angeles, California

John & Mary Markle Foundation
New York, New York

Richard King Mellon Foundation
Pittsburgh, Pennsylvania

Eugene & Agnes E. Meyer Foundation
Washington, D.C.

Charles Stewart Mott Foundation
Flint, Michigan

New York Community Trust
New York, New York

Powell Family Foundation
Overland Park, Kansas

Charles H. Revson Foundation
New York, New York

Russell Sage Foundation
New York, New York

St. Paul Community Foundation
St. Paul, Minnesota

None of these individuals or organizations bear ultimate responsibility for any errors of fact or interpretation that remain here. That responsibility is ours alone.

L.M.S.
K.P.V.

About The Authors

LESTER M. SALAMON was born in Pittsburgh, Pennsylvania. He earned a B.A. in Economics and Public Policy from Princeton University (Phi Beta Kappa) in 1964 and a Ph.D. in Government from Harvard University in 1971.

Dr. Salamon is currently Professor of Political Science and Director of the Institute for Policy Studies, The Johns Hopkins University. Prior to this he was Director, Center for Governance and Management Research, The Urban Institute (1980-1986); Deputy Associate Director, U.S. Office of Management and Budget (1977-1979); Associate Professor of Policy Sciences and Political Science and Director of the Center for Urban and Regional Development Policy, Duke University (1973-1980, on leave 1977-1980); Assistant Professor of Political Science, Vanderbilt University (1970-1973); and Instructor, Department of Political Science, Tougaloo College (1966-1967).

In addition to these professional positions, Dr. Salamon has served as Consultant to the National Assembly of Voluntary Health and Human Services Agencies, the San Francisco Foundation, United Way of the Bay Area; and Senior Consultant to the Presidential Management Panel, National Academy of Public Administration. He was also a Ford Foundation Associate; Director, Duke-OMBE Land Project; Associate, Brookings Institution; Director, Banking Committee Study, Ralph Nader Congress Project; Associate, Abt Associates, Inc.; Associate, Ralph Nader's Center for Study of Responsive Law; Research Associate, Hudson Institute; and Management Intern, Office of the Secretary of Defense.

Dr. Salamon is on the board or a member of a number of professional associations and the winner of many academic awards and fellowships. He is a frequent lecturer both in the U.S. and abroad and has authored numerous books, monographs, and articles on government policy and the nonprofit sector. His most recent book is *Beyond Privatization: The Tools of Government Action* (The Urban Institute Press, 1989).

KENNETH P. VOYTEK earned a B.A. in Economics from Saint Edwards University in 1980 and an M.A. in Public Policy from West Virginia University in 1985.

He is currently an Economist with the Center for Local Economic Competitiveness, Michigan Department of Commerce, working on the collection and analysis of substate economic data and strategic economic development reports. During 1984-1987, Mr. Voytek was a Research Associate with The Urban Institute where he co-directed the Foundation Fiscal Policies and Practices Study. In addition, he was responsible for the development of an analytical framework to help state government officials evaluate the potential for using the private sector to deliver state government services.

Mr. Voytek's other professional positions have included Graduate Assistant, Department of Political Science, West Virginia Univerity (1983-1984), and Research Associate, Center for Governmental Studies, Northern Illinois University (1981-1983), where he was responsible for developing and directing a research agenda for state and local economic development policy and providing technical assistance to state, county, and local governments.

Mr. Voytek is author of numerous reports and papers on state and local economic policies and performance and was named an Outstanding Young Man of America in 1986.

MANAGING FOUNDATION ASSETS

by
Lester M. Salamon and Kenneth P. Voytek

EXECUTIVE SUMMARY

BACKGROUND

The more than 25,000 private, charitable foundations that exist in the United States provide a unique source of stable support for the nation's private, non-profit sector.

To function effectively, however, foundations must not only make sensible decisions about the kinds of projects and organizations to support; they must also successfully manage the substantial assets in their control. This requires balancing the needs of current versus future beneficiaries in a context of considerable uncertainty and risk.

Prior to 1969, foundations were largely on their own in making these judgments, albeit within a context of legislative warnings against the use of the foundation device as a tax dodge. With the passage of the Tax Reform Act of 1969, however, foundations were subjected to a variety of more specific regulations on their investment and "payout" practices. Most significantly, they were required to pay out in "qualifying distributions" (i.e., grants plus eligible program expenses) the higher of (a) a given percentage (ultimately, 5 percent) of their assets, or (b) their total investment earnings in a given year. To maintain an asset base capable of supporting grantmaking into the future, foundations therefore had to generate a total return on their investments at least equal to the inflation rate plus 5 percent.

As it turned out, this "payout requirement" imposed a significant burden on foundations during the high-inflation years of the 1970s. Responding to foundation complaints, Congress agreed in the 1981 tax act to liberalize the payout requirement by eliminating the requirement that foundations must pay out the full value of their investment earnings if these exceeded 5 percent of asset value. Henceforth, foundations could retain earnings in excess of this amount and reinvest them to maintain or increase the real value of their asset base. In making this move, however, Congress expressed interest in knowing what effect it would have on both foundation investment and payout performance.

OBJECTIVES OF THIS REPORT

Unfortunately, the data needed to provide such an assessment have long been lacking. Despite their importance to the functioning of foundations and the attention they have attracted in Congress, foundation investment and payout management have received precious little research attention. As a consequence, little is known in solid empirical terms about how foundations manage their investments and grant decisions or what their actual performance has been. What limited work has been done on this topic has either focused on too narrow a range of foundations, too limited a definition of investment performance, or too restricted a time frame to instill much confidence in the results.

To fill this gap in existing knowledge and begin the task of developing a solid body of empirical data on foundation investment and payout activities, the Council on Foundations commissioned this inquiry on foundation fiscal policies and practices. In particular, this project sought to answer five principal questions:

How are foundation investment decisions made? What investment policies and objectives are foundations pursuing? What variations in policies exist and what accounts for them?

1

What rates of return have foundations achieved on their investments? What variations in investment performance exist and what accounts for them?

How are foundation payout decisions made? What policies or guidelines shape these decisions?

What have been the payout rates of foundations? What accounts for the variations that exist?

What impact has the change in the payout requirement in 1981 had on either foundation investment and payout policy, or foundation investment and payout performance?

APPROACH

To answer these questions, two new bodies of data were developed.

First, a survey was administered by mail to a stratified random sample of over 1,000 private foundations throughout the country. This survey sought information primarily on how foundations manage their investment and payout processes, and on the investment and payout policies they are pursuing. A stratified random sample was used in view of the uneven distribution of foundation assets: less than 1 percent of all foundations control over half of all foundation assets. In order to ensure sufficient representation of these large foundations in the final sample, it was necessary to "over-sample" them. Altogether, 478 foundations provided complete and valid responses to this survey. The result is a rich new source of information on foundation fiscal practices.

Second, a sizable body of financial data was assembled from the 990-PF forms that foundations are required to file each year with the Internal Revenue Service. These data made it possible to analyze actual investment and payout performance. Form 990-PF data were sought on all foundations with assets in excess of $10 million and all foundations with assets under $10 million that responded to our survey. Altogether, data were available on 459 such foundations over a five-year time span, 1979–1983.

Data from both the survey and the 990-PF data collection were then "blown up" to the total population using two weighting schemes—one to represent the universe of foundations and the other to approximate the universe of foundation assets.

In order to put foundation performance into context and hold constant the impact of market conditions on the results, we also collected data on the performance of a comparable "control portfolio" composed of 60 percent common stock, 30 percent fixed-income securities, and 10 percent cash or cash equivalents (e.g., Treasury bills). We assumed that this control portfolio performed the way standard stock, bond, and Treasury bill indexes, respectively, performed during this time period.

CAVEATS

The following discussion summarizes the major findings of this work. In the interpretation of these findings, two important considerations should be kept in mind.

First, the time period covered in most of this report, 1979–1983, was an unusually favorable one in the capital markets, but it followed an unusually unfavorable one. It was also a rather volatile period during which interest rates and the inflation rate hit historic highs. While the use of a control group allows us to factor out some of the effects of the time period, it is not possible to factor out all of these effects.

Second, investment management inevitably involves a trade-off between risk and return. Generally speaking, in order to have a reasonable prospect of higher returns, it is necessary to take greater risks and to be willing to absorb greater volatility. Lower returns can therefore be a rational option when the alternative is to put one's assets at greater risk or when one wants to maximize payout in the short run and therefore must choose investments that achieve higher current earnings at the expense of long-term growth.

Against this backdrop, what have we learned about foundation investment and payout processes, policies, and performance?

FINDINGS

Investment Management Processes and Policies
(Chapters Two and Three)

1. Two broad patterns of investment management seem to exist in the foundation community. The first of these resembles commonly recommended investment practice and features the active involvement of foundation boards in setting broad investment objectives, an active management style involving regular scrutiny of performance and asset composition, a focus on "total return" (i.e., income plus appreciation) as the investment objective rather than just investment income alone, a level of risk tolerance comparable to that associated with standard stock indexes, a three-to-five-year time horizon, use of multiple outside investment managers, and moderate rate of return goals. The second features limited board involvement in setting investment objectives, a hands-off management style, a low level of risk tolerance, preoccupation with income instead of total return, a one-year investment time horizon, and limited rate-of-return goals.

2. The first of these two patterns is most common among the larger foundations, which control most of the foundation assets. However, about half of the large foundations, about 80 percent of the moderate-sized foundations, and about 95 percent of the small foundations utilize an approach that is much closer to the second one identified above. What this means is that while a significant share of foundation assets is managed in the more active and aggressive style, most foundations pursue a much less active and aggressive approach, with far less involvement by foundation boards, little tolerance for risk, and limited goals and time horizons.

3. This disparity finds expression in the composition of foundation investment assets. Close to 60 percent of all foundation assets are held in the form of equities. However, this largely reflects the asset composition of the larger foundations. Most foundations are more heavily invested in fixed-income securities.

4. There is some evidence that the 1981 change in the payout requirement has helped to encourage a more flexible investment approach among foundations, with more emphasis on total return rather than income yield.

Investment Performance
(Chapter Four)

5. The investment performance at least of the larger foundations was quite respectable during 1979–1983. In fact, the average annual rate of total return (income plus appreciation or depreciation) on foundation assets during 1979–1983 exceeded the rate on a comparable control portfolio (14.1 percent versus 13.5 percent). After adjustment for inflation, the rate of return on foundation assets was close to 6 percent a year during 1979–1983, which was enough to support a 5-percent-payout rate without eating into the real value of the asset base.

6. This investment record was largely the product of the superior performance of the larger foundations, however. By contrast, most foundations had investment performance that was below the control portfolio. In particular, the average annual rate of total return for the median foundation represented in our sample was 12.4 percent. After adjustment for inflation, this translates into a real rate of growth of only 4 percent a year—not enough to sustain the real value of the asset base and still support a 5-percent-payout rate. The real value of the assets of most foundations therefore suffered a decline during this period despite the generally favorable investment climate that existed.

7. While the large foundations generally had higher total return rates than the smaller foundations during this period, the smaller foundations generally had higher income yield. This likely reflects the greater tendency of the smaller foundations to invest in fixed-income securities and their evident preference for less risk and volatility.

8. Foundation investment performance was generally better during the latter part of this period (after the change in the payout requirement) than it was during the earlier part (prior to the change in the payout requirement). However, this was probably due more to a dramatic improvement in market conditions than to the impact of the payout requirement change. Significantly, foundations did not benefit as much from the improvement in market conditions as did the control portfolio. This probably reflects the somewhat more conservative investment approach of foundations, which allows them to do better than the market indexes during weak periods but prevents them from doing as well as the market indexes during strong periods.

9. Among the factors that account for variations in foundation investment performance, the most significant appear to be the degree of activism in the management of assets and the composition of the investment portfolio. However, both of these are essentially measures of the degree of risk a foundation is willing to sustain. At least during the period covered here, greater tolerance for risk brought with it greater investment returns. This is consistent with modern investment theory, but it may provide little help to those foundations that prefer to accept lower total returns in order to avoid uncomfortable levels of risk.

Payout Policy and Performance
(Chapter Five)

10. Compared to the minimum 5-percent-payout rate mandated by law, the median foundation represented in our sample registered a payout rate of 7.8 percent a year during the 1979–1983 period.

11. Generally speaking, payout rates tended to be higher among the smaller foundations than among the larger ones. Two factors appear to be responsible for this: first, differences in payout policy, and second, differences in investment strategy.

As far as payout policy is concerned, the smaller foundations are much more likely to have their payout rate dictated by their investment performance and to pay out all of their investment income. By contrast, the larger foundations are more likely to pay out the minimum that the law requires, or slightly above this, and to orient their investments in such a way as to permit this. In other words, for the former group of foundations, investment decisions drive payout decisions, whereas for the latter, payout considerations influence investment activities.

The investment strategies of the two groups of foundations reflect this. Generally speaking, the larger foundations were more heavily invested in common stocks, which tend to have lower yields but greater potential for growth. This permitted these foundations to keep their payout rates close to the 5-percent minimum. By contrast, the smaller foundations tended to focus more heavily on fixed-income securities, which had higher income yields during this period and therefore required higher payout.

12. The foundations that stand to benefit the most from the 1981 change in the payout requirement are therefore paradoxically the ones with the highest income yield but the lowest rates of total return (income plus appreciation). By virtue of their high income yield, these are the foundations that generally had payout levels in excess of the 5-percent minimum and that therefore have had the option since 1981 of not paying out this "surplus." Thus, it becomes especially important for the foundation community to encourage these foundations to make sensible and effective decisions about how to utilize the savings that may result.

13. There was little evidence through 1984 of a significant change in the payout rate in response to the 1981 change in the payout requirement. Although the payout rate declined in 1984, this was probably more a product of a decline in income yields than of any change in payout activity. This seems to confirm the notion that the major beneficiaries of the payout change are the less aggressive foundations, which tend to invest more heavily in fixed-income securities and to pay out all the earnings that result, payout requirement or not.

Recommendations
(Chapter Six)

Extension of 1981 Payout Law Change

The data in this report provide some confirmation of the wisdom of the liberalization of the payout requirement in 1981. This change has allowed some greater flexibility for foundations in designing their investment strategies without evident negative side-effects in terms of foundation payout rates. The 1981 change should therefore be sustained.

Multiyear Basis for Payout Computation

While the 1981 change in the payout requirement was a step in the right direction, it may not have gone far enough. A useful next step would be to replace the current one-year base for computing the payout rate with a three- or five-year floating average. This would allow foundations to lengthen their investment time horizons and also reduce the year-to-year fluctuations in required distributions, which can disrupt foundation program planning.

More Attention to Foundation Investment Management

In view of the fact that most foundations seem to be pursuing rather ad hoc investment approaches and achieving rates of return below the market averages for a comparable portfolio, more attention probably needs to be focused on the investment function of foundations. This could take the form of training sessions and educational materials as well as a more systematic and permanent mechanism for tracking foundation investment behavior.

A Common Fund for Foundations

To equip the smaller foundations in particular to operate as successfully as possible in the highly specialized capital markets, the foundation community should seriously consider borrowing a page from the higher education field and organize a "common fund" for foundations. Such a fund could pool the investment resources of a number of foundations and manage them centrally, permitting small- and medium-sized foundations to benefit from expert management at lower cost than is now possible.

Conclusion

In the wake of government budget cuts and continued constraints on government spending, private foundations have gained unexpected prominence in American life. In this context, it is critically important to continue the improvement in foundation investment behavior that is already underway and extend it to the considerable number of foundations that still seem to be lagging behind.

By demonstrating that a significant number of foundations are performing quite well while others still have some distance to go, we hope that this report will both provide a basis for protecting the gains that have been made and a stimulus for encouraging the changes that are needed. Foundation investment management is too important to the fiscal health of the nonprofit sector and, through it, to the social and political health of the nation, to remain hidden from view. It must now be given the sustained attention it deserves.

Chapter One

INTRODUCTION

Almost 25,000 private, charitable foundations exist in the United States today. With assets of $92.6 billion as of 1985, these institutions represent one of the few solid sources of financial certainty for the nation's private, nonprofit sector. In 1985 alone, foundations generated nearly $6 billion in grants for the nations' nonprofit organizations.[1] By comparison, United Way, the workplace fundraising campaign that is the largest single source of individual donations to charitable activities, generated just over $2 billion, or about a third as much support, for such organizations that same year.

More important than their size, however, foundation grants provide both stability and flexibility. Foundation support derives not from contributions generated in a particular year but from the proceeds of endowments contributed in the past. As a result, it is both more certain and potentially more resistant to outside pressures. Hence, although the financial support provided by foundations constitutes less than 5 percent of the total income of the nation's nonprofit organizations,[2] it plays a special role in the life of the sector by providing a durable base of support and performing a potentially "re-

newing" function through support for the unknown or the controversial.[3]

For foundations to perform this role, however, they must make sensible decisions about the resources entrusted to their care. This involves more than the selection of effective projects to support; it also involves the careful management of the assets in foundation control.

In a sense, foundations are called upon to balance two potentially competing goals: first, the promotion of charitable good deeds through the distribution of grants, and second, the preservation and enlargement of the financial base needed to support this grant-making activity into the future.

Although a small number of foundations function as "pass-through" organizations that essentially collect funds from donors and distribute them within the same year as grants to nonprofit organizations, most foundations have assets that they must manage over time. Decisions therefore have to be made about how to invest these assets and what portion of the earnings or principal should be paid out in grants each year.

[3] On the special role of private foundations in American society, see: Ben Whitaker, *The Philanthropoids* (New York: William Morrow and Company, 1974), pp. 167–194; Council on Foundations, "Private Foundations and the 1969 Tax Reform Act," in *Research Papers*, vol. III (Washington, D.C.: Commission on Private Philanthropy and Public Needs, 1977), pp. 1576–1584; *Foundations, Private Giving and Public Policy: Report and Recommendations of the Commission on Foundations and Private Philanthropy* (Chicago: The University of Chicago Press, 1970), pp. 39–53, 93–111.

[1] *The Foundation Directory*, 11th ed. (New York: The Foundation Center, 1987), p. xiv.

[2] Based on data in Lester M. Salamon and Alan J. Abramson, *The Federal Budget and the Nonprofit Sector* (Washington, D.C.: The Urban Institute Press, 1982).

If foundations pay out more than they earn, over time their assets will decline, and they will cease to exist. On the other hand, if they retain too large a share of their earnings rather than distribute them for charitable purposes, legitimate questions will arise about the wisdom of granting them tax-exempt status. Foundations can respond to this dilemma by seeking higher returns on their investments, thus supporting both higher charitable expenditures and greater asset growth, but this course often involves risks that could jeopardize both expenditures and asset growth in future years. Foundations must therefore negotiate a difficult course between the competing demands of current and future claimants in a context that inevitably involves uncertainty and risk.

I.
THE ISSUE: AN APPROPRIATE BALANCE?

Unfortunately, concerns have arisen in the past about the way in which foundations were negotiating this course, and particularly about whether they were maintaining an appropriate balance between their investment and grant-making functions. Celebrated instances of apparent misuse of the foundation device gave rise during the 1950s and 1960s to fears that the private foundation was being transformed into a tax dodge allowing wealthy families to evade the inheritance laws while retaining control of their businesses. In fact, financial advisors promoted the foundation device during the 1940s and 1950s in precisely these terms, pointing out that families could transfer their businesses to a private foundation without paying taxes, name family members as trustees, and then use the resulting instrumentality to run the business and accumulate profits without distributing the proceeds to charity.[4]

Inquiries launched by Congress and by the U.S. Treasury Department in the 1950s and 1960s demonstrated that at least some foundations were operating in precisely this fashion.[5] This seems to be confirmed by a recent analysis of foundation growth patterns, which attributes the surge in foundation formations that occurred in the 1940s, at least in part, to the financial advantages foundations were felt to provide to wealthy donors as a consequence.[6]

Responding to such disclosures, Congress in 1950 passed the first in what would be a series of restrictions on private foundation operations. In the Revenue Act of 1950, Congress threatened the tax exemption of any nonprofit organization that was involved in "self-dealing" or that made "unreasonable accumulations of income."[7] The definitions of these terms were left extremely vague, however. What is more, the penalty for failure to comply with the law was so severe—complete loss of tax-exempt status—that enforcement lagged. When problems persisted, Congress therefore responded with additional restraints that culminated in the Tax Reform Act of 1969.

In addition to singling foundations out from other tax-exempt organizations for the first time, the 1969 Tax Act established an elaborate set of restrictions on foundation investment and grant-making behavior.[8] In the first place, in order to deter "self-dealing" between the foundation and others with a stake in its operation, the act imposed a 5-percent excise tax on financial transactions between foundations and certain "disqualified persons," which included contributors to the foundation, the foundation's managers and trustees, certain public officials, and family members of disqualified persons. Second, to deter use of the foundation device to retain control of family-owned businesses, the act also imposed a 5-percent fine if a foundation or any of its "disqualified persons" holds more than a combined 20-percent interest in any business. Third, foundations were subjected to a 5-percent penalty tax on so-called jeopardy investments, i.e., investments that violated ordinary standards of business prudence and thereby jeopardized the charitable purpose of the foundation. Fourth, certain types of foundation grants—e.g., those intended to influence legislation, to intervene in politi-

---

[4] "How to Have Your Own Foundation: Taxation Has Brought the Charitable Instrument of the Rockefellers and Carnegies Within the Reach of Thousands," *Fortune* (August 1947).

[5] The most notable of these hearings were conducted by the Committee on Small Business of the U.S. House of Representatives under the chairmanship of Congressman Wright Patman (Dem., Tex.). In a series of reports, the Patman Committee documented numerous instances of apparent use of the foundation device primarily for personal and family financial gain. Some of the Patman findings were later confirmed in a report prepared by the U.S. Treasury Department in 1965. See: U.S. House, Committee on Ways and Means, *Treasury Department Report on Private Foundations,* 89th Cong., 1st sess., (1965), pp. iii, 1–2.

[6] Gabriel Rudney, "The Creation of Foundations and Their Wealth," in Teresa Odendahl, ed., *America's Wealthy and the Future of Foundations* (New York: The Foundation Center, 1987), pp. 190–192.

[7] Revenue Act of 1950, ch. 994, sec. 331, 64 Stat. 906, 957.

[8] For more detailed analyses of the provisions of the 1969 Tax Act pertaining to private foundations, see: U.S., Committee on Ways and Means, Subcommittee on Oversight, *Development of the Law and Continuing Issues in the Tax Treatment of Private Foundations,* June 17, 1983; John Edie, "Congress and Foundations: Historical Summary," in Teresa Odendahl, ed. *America's Wealthy and the Future of Foundations* (New York: The Foundation Center, 1987); David F. Freeman, *The Handbook on Private Foundations* (Cabin John, Md.: Seven Locks Press, 1981), pp. 59–72 and Appendix 9.

cal campaigns, to provide support for individuals—were to be penalized as "taxable expenditures" unless they fell within certain enumerated exceptions or were handled with special care.

Finally, and perhaps most important, the 1969 law established a "minimum payout requirement" to deter foundations from accumulating resources and delaying their distribution for valid charitable purposes. Foundations were required to distribute each year either the full amount of their investment income or a given percentage—initially set at 6 percent—of their investment assets, whichever was higher.[9] This payout requirement was designed to prevent foundations from functioning as essentially tax-free investment pools. Henceforth, a foundation's investment performance would be tied much more closely to its charitable activities and vice versa.

Although the 1969 act established a variety of "transition" provisions to phase in the new restrictions, it nevertheless marked a major turning point in the operation of private foundations. Henceforth, the foundation investment function and the link between investment and grant making took on even greater importance, not only as crucial aspects of foundation management, but also as the focal point of governmental involvement in the foundation field.

II.
THE PROBLEM: THE LACK OF DATA

Despite the legislative attention they attracted and their obvious importance to the operation of foundations, however, foundation investment management and the relationship between investment performance and charitable activity have remained largely unexplored in serious empirical terms. Indeed, the decision to subject foundations to a comprehensive regulatory regime in the 1969 tax act was based largely on anecdotal evidence rather than on a broad-based empirical inquiry into the investment and payout performance of a systematic cross-section of foundations. The ink was no sooner dry on the law, therefore, than three serious criticisms surfaced about its likely consequences.

The first criticism charged the 1969 act with threatening to erode the value of foundation assets by forcing

a higher payout rate than could reasonably be sustained given prevailing inflation rates and market conditions. In a celebrated 1976 article, for example, Professor J. Peter Williamson of Dartmouth College pointed out that the rate of return foundations could expect from a prudently managed fund yielding returns in line with the recent performance of stock and bond indexes would not be sufficient to support a payout rate of 5 or 6 percent of the foundation's assets and still keep pace with inflation. The effect of the payout requirement, Williamson argued, would thus be to reduce the real value of foundation assets over time.[10] In a companion piece written five years later, Williamson argued that, if anything, his earlier estimates about the impact of the minimum payout requirement had been overly optimistic in view of the inflation record of the late 1970s.[11]

A second criticism focused on the impact of the 1969 payout requirement on the management of the foundation grant-making function. The argument here was that the law imposed an excessive degree of variability on the grant-making activities of foundations. This variability was partly a product of the original requirement that the minimum payout rate was to vary with changes in certain economic indicators (such as interest rates). But it also reflected the fact that the requirement to pay out all earnings, or a given percentage of assets, in a particular year made the grant-making activities of foundations dependent on the sometimes whimsical movements of securities markets, complicating the development of coherent charitable programs.

A third criticism faulted the 1969 law for unfairly constraining the investment behavior of foundations and at times inducing investment decisions that were unwise from the perspective of the long-term viability of foundation assets. Depending on the relationship between interest rates and inflation rates, foundation investment managers could find themselves pushed by the payout requirement into heavier investments in long-term bonds or low-yield securities than they might consider appropriate. When interest rates exceeded the minimum payout rate, for example, the requirement in the 1969 law that foundations pay out all of their investment earnings created strong incentives for foundation investment managers to invest in securities that have lower yields but greater prospects for appreciation in value. The problem, however, is that such investments also involve greater risk of price volatility. As a consequence, the regulation may have tended to make

---

[9] Investment income includes dividends, interest, and other annual earnings on investments but does not include unrealized gains or losses in the value of assets. The assets against which the minimum percentage is applied is a monthly average of the market value of the foundation's assets. The law also makes provision for foundations to "carry over" payout levels in excess of the minimum in any year to any of the subsequent five years.

[10] J. Peter Williamson, "Investment Expectations and the Foundation Payout Rate," *Foundation News* (January/February 1976), pp. 13–18.

[11] J. Peter Williamson, "Inflation and the Foundation Payout Rate," *Foundation News* (March/April 1981), pp. 18–24.

foundation portfolios more volatile than they would have been otherwise.[12]

In the years since 1969, Congress has responded to at least some of these arguments with changes in the 1969 payout requirement.

In 1976 the minimum investment return requirement—the percentage of assets to be paid out—was changed from a variable rate to a fixed rate of 5 percent of asset value. Then, in 1981, Congress made a more fundamental change: It removed the stipulation that foundations had to pay out any investment income in excess of 5 percent of asset value. Henceforth, payouts in excess of 5 percent were to be at the discretion of the foundation. This change meant that foundations could now choose to retain investment earnings in excess of 5 percent of asset value, using that excess to enlarge the asset base or to keep pace with inflation.

In theory, at least, these changes have freed foundations to manage their investments more flexibly and to achieve a better balance between current payout and future payout capacity. But will the changes achieve these objectives in reality? The risk of another round of regulatory impositions depends significantly on the answer to this question. Unfortunately, however, no mechanism is ready at hand for gauging this. Almost two decades after the enactment of the 1969 law, and a quarter of a century after Congress first imposed restrictions on foundation investment and payout activities, no solid body of data exists on how foundations manage their assets, how well they have been performing, or how investment performance relates to payout performance.

The few analyses of foundation investment activity that have been undertaken have been narrowly focused or limited in coverage. A major study of foundation investment practices conducted by Professor Ralph Nelson in the mid-1960s, for example, focused chiefly on the *composition* of investment portfolios and the process of diversification. Although Nelson attempted some computation of total investment return, he covered only 45 very large foundations.[13] Similarly, in the late 1960s the Commission on Foundations and Private Philanthropy (the Peterson Commission) computed the return rate for some 180 foundations, but the analysis covered only one year (1968), and it is difficult to determine what methodology the commission used.[14] Indeed, the

Commission acknowledged in its report that it made "no claims that it has exhaustively reviewed the investment performance of foundations."[15]

When Peter Williamson prepared his analyses of the likely effect of the 1969 foundation payout requirement in the mid- and late-1970s, therefore, he had to rely on simulations rather than actual foundation investment data. Indeed, Williamson acknowledged that not only were data lacking on the actual investment performance of foundations, but also "[t]here appear to be no readily available statistics on the composition of private foundation investment portfolios."[16]

Three more recent studies have shed additional, useful light on foundation investment activities, but none of these provides the kind of broad-gauged empirical picture that is needed. The first of these, completed by researchers at the University of Michigan Graduate School of Business Administration, covers only 39 foundations and primarily focuses on the investment management *process* and how it was affected by the 1981 change in the payout requirement, with no attempt to calculate actual foundation investment-return rates.[17] A far more ambitious analysis of the investment activities of a large sample of 1,309 foundations was completed by the Statistics of Income Division of the Internal Revenue Service in 1985, but this study focused only on income yield rather than total rate of return, examined only a single year, and made no adjustment for the widely different fiscal years of foundations.[18] Finally, Professor Ralph Nelson has recently completed an analysis of the growth of private foundation assets between 1962 and 1982, but this analysis develops rate-of-return figures on only the 54 largest foundations for which detailed financial profiles could be developed over this 20-year period.[19]

Nor do the commercial investment reporting services provide much help. Although some foundations are included in the data generated by some of these services, such as Colonial Consulting and Hamilton and Johnson, the coverage is spotty. Foundations essentially self-

[12] Williamson, "Inflation and the Foundation Payout Rate," 1981, pp. 20–24.

[13] Ralph Nelson, *The Investment Policies of Private Foundations* (New York: Russell Sage Foundation, 1967), especially pp. 70–77.

[14] Commission on Foundations and Private Philanthropy, *Foundations, Private Giving, and Public Policy* (Chicago: University of Chicago Press, 1970), pp. 72–75 (cited hereafter as *Peterson Report*).

[15] *Peterson Report*, p. 75. The Commission did report, however, that "all the signs it has seen tend to say the same thing: that the investment performance of foundations is below par, and perhaps significantly so" (p. 75).

[16] Williamson, "Inflation and the Foundation Payout Rate," p. 19.

[17] Raymond R. Reilly and Donald H. Skadden, *Private Foundations: The Payout Requirement and its Effect on Investment Spending Policy* (Ann Arbor: University of Michigan Graduate School of Business Administration, 1981).

[18] Margaret Riley, "Private Foundation Information Returns, 1982," *Exempt Organization Studies* (Statistics of Income Division, U.S. Internal Revenue Service, July 1985).

[19] Ralph Nelson, "An Economic History of Large Foundations," in Teresa Odendahl, ed., *America's Wealthy and the Future of Foundations,* (New York: The Foundation Center, 1987), pp. 127–177, especially pp. 155–160.

select themselves into the services, making the resulting data suspect as a basis for generalizing about the foundation universe as a whole. What is more, the number of foundations covered is small—less than 50 for Colonial Consulting and not much more for Hamilton and Johnson.[20]

In short, the available data on foundation investment management and on the relationship between investment performance and payout activity remain almost as sparse today as they were in 1969 when Congress enacted the original payout requirement. This lack of good information creates a danger that assessments of the consequences of the change in the payout requirement enacted in 1981 will be based, like the 1969 law, more on isolated anecdotes than on solid empirical data reflecting the experiences of a reasonable cross-section of foundations. Furthermore, the foundation community may continue to miss important opportunities to improve the overall investment performance of foundations, and thus limit the growth of resources available for support of charitable activities.

III.
PURPOSE OF THIS STUDY

It was to help fill this gap in knowledge and create a solid body of empirical data on foundation investment policies and practices that the Council on Foundations launched the inquiry reported on here. In particular, this project had three principal objectives:

- *First,* to examine foundation investment *practices* and *policies*—the roles played by various participants (foundation trustees, officers, outside managers, and staff), the content of foundation investment objectives and strategies, and the relationships between investment and payout decisions.

- *Second,* to determine the actual investment and payout *performance* of foundations and explain what accounts for the patterns of variation that are apparent.

- *Third,* to assess what impact, if any, the change in the foundation payout requirement in 1981 had either on the management of the foundation investment and payout processes, or on foundation investment and payout performance.

In the pursuit of these three objectives, moreover, four additional considerations were also important:

- *First,* a desire to cover the entire foundation universe, and not just foundations in one size class or

[20] A third investment reporting service, InData, covers 157 institutions in its endowment fund universe, but many of these are universities and other non-foundation endowments, and all 157 have assets of only $5.3 billion compared to the $92.6 billion in foundation assets.

region of the country. This was important because prior work had tended to focus on only the largest foundations, those with assets in excess of $50 million. While these foundations account for 55 percent of foundation assets, they represent less than 1 percent of all foundations (see Table 1.1). Our goal was to be able to make generalizations about the investment performance of both large and small foundations and thus to support observations both about the universe of foundation *assets* and the universe of *foundations.*

Table 1.1
Distribution of Foundations and Foundation Assets, by Size Class, 1982

Asset Size Class	Percent of Foundations[a] (N = 22,299)	Percent of Assets[a] (N = $47 billion)
$50 million and over	0.7	55.1
$10–49.9 million	2.5	20.8
$1–9.9 million	14.8	18.0
Under $1 million	82.0	6.1
Total	100.0	100.0

Source: National Data Book, 9th Ed. (New York: The Foundation Center, 1985), p. x.

[a] Does not include 1,309 "operating" foundations with assets of $2.5 billion that did not award any grants in the current year.

- *Second,* a preference for hard, empirical data, rather than subjective judgments, simulations, or anecdotes.

- *Third,* the use of a reasonable time period in order to cancel out the impact of short-term fluctuations and to take account of variations in the definition of foundation fiscal years.

- *Fourth,* a determination to factor out the impact of external influences such as market conditions to the extent possible by developing data on a reasonable control group against which foundation performance could meaningfully be compared.

IV.
METHODOLOGY AND APPROACH

To pursue these objectives consistently with the considerations outlined above, we developed a project that involved five major features: first, a *mail survey* of a broad cross-section of private foundations; second, the collection of detailed financial data from the *990-PF forms* foundations submit each year to the Internal Revenue Service; third, focus on a *five-year time span,* 1979–1983; fourth, the use of a *control group* to compare the performance of foundations to that of other

Table 1.2
Mail Survey Sample and Response Rates

Asset Size	Number of Foundations[a]	Number Surveyed	Target Sample	Actual Sample	Actual Response as Percent of Target
Over $50 million	166	166	128	74	57.8
$10–49.9 million	533	533	213	124	58.2
Below $10 million	21,595	4,649	822	280	34.1
Total	22,294	5,348	1,163	478	41.1

Source: Salamon/Voytek Foundation Investment Study.

[a] Excludes company-sponsored and community foundations.

investors and thus filter out the effects of the particular time period covered; and fifth, two *weighting schemes* to blow the sample up, respectively, to the universe of foundations and the universe of foundation assets. The discussion below briefly discusses each of these features.[21]

Mail Survey of Foundations

To gather information on foundation investment practices and policies, we circulated a mail survey to a sample of private foundations selected randomly from lists provided by the Foundation Center. The survey form was developed in consultation with a project advisory committee and then field-tested on a random sample of 30 foundations. Based on this field test, we simplified the survey form, submitted it to our advisory committee for review, and then distributed it to the full sample of foundations in October 1985.

Because of the skewed distribution of assets within the foundation universe, we could not rely on a straight random sample for this survey. With 55 percent of all foundation assets under the control of less than 1 percent of the foundations (see Table 1.1), such a straight random sample would have included too few of the 166 largest foundations and thus would have underrepresented grossly the vast majority of foundation assets.

Accordingly, we utilized a disproportionate, stratified random sample and oversampled the largest foundations. We divided the foundation universe into four size classes based on the 1982 assets of the foundations. Survey forms were then distributed to all 699 of the foundations in the two largest size classes (i.e., all foundations with assets of $10 million or more), to one-third of the foundations with assets of $1 million to $9.9 million, and to 20 percent of the foundations with assets under $1 million. This yielded a potential sample of 5,348, from which, based on previous experience,

we aimed at an actual target sample of approximately 1,100 (See Table 1.2).

As it turned out, 478 usable surveys were returned, or slightly better than 40 percent of the target number.[22] Among the larger foundations, the return rate was close to two-thirds, which is extraordinary for work of this sort. Altogether, the foundations in the sample represent 2 percent of all foundations and about 40 percent of all foundation assets. Most important, the return rate was sufficient in each of the size classes separately to support valid inferences both about each of the size classes and about the foundation universe as a whole.[23]

Form 990-PF Data Collection

Although we originally contemplated collecting investment data, and not just information on the investment process, through the mail survey form, the pretest experience made clear that this was an uneven and imperfect way to collect such empirical data. Few foundations completed the pretest survey forms completely, and the data provided left too many questions. Foundations reported that historical data on portfolio composition or investment performance going back even a few years were essentially unavailable. In addition, significant disparities existed in the methods different foundations use for computing return rates and in the way they group their assets in their reports (e.g., some grouped corporate and government bonds together and reported equities [stock] separately while others put

[21] For additional details on our methodology, see Appendix A.

[22] A parallel survey was conducted of community foundations, yielding 49 valid returns. However, the results reported here do not include these community foundations.

[23] For a discussion of disproportionate stratified random sampling, see: Leslie Kish, *Survey Sampling* (New York: John Wiley and Sons, Inc., 1967), and Charles H. Backstrom and Gerald Hursh-Cesar, *Survey Research* (New York: John Wiley and Sons, Inc., 1981). For a discussion of the standard error of the estimate associated with this sample, see Appendix A. For the sample as a whole, we are 95 percent certain that the confidence interval ranges between ±1.0 and ±4.5 percent. Confidence intervals for the sub-samples are slightly higher.

corporate stock and bonds together and reported government securities separately).

Rather than rely on foundation self-reporting through the mail survey form, therefore, we turned to the financial information that foundations provide to the Internal Revenue Service on their annual Form 990-PF filings. These 990-PF forms contain a vast amount of detailed financial data on foundations. What is more, the forms are regularly copied by the Internal Revenue Service and provided in microfiche form to the Foundation Center, which stores them both in its Washington and New York offices. While the 990-PF data forms are not intended to report on investment return rates, we found that by using certain assumptions (which are spelled out in more detail in Chapter Three), we could utilize the data on these forms to develop a good approximation of foundation investment performance.

Accordingly, we coded a large body of investment data from the 990-PF forms stored at the Foundation Center for a sample of foundations for the years 1979–1983. In addition, in order to include data covering three years on either side of the 1981 change in the payout requirement, we also asked the survey sample of foundations to provide us with copies of their 1984 forms, which were not yet available at the Foundation Center when we were conducting our research.

The sampling approach used for this portion of the analysis was similar to the disproportionate, stratified sampling method used for the survey. In particular, we sought 990-PF data on all foundations with assets in excess of $10 million, and on all 304 foundations with assets under $10 million that responded to our mail survey. This was done to enable us to link the investment data to the survey results. Altogether, 990-PF data were sought on 1,003 foundations. Ultimately, complete data files covering all five years were assembled on 459 of these foundations.

The data set resulting from this 990-PF data collection effort is extraordinarily rich. At the same time, however, like any empirical data set, this one also has several limitations which should be kept in mind in interpreting the results, and which are described in the following paragraphs.

Asset Composition

In the first place, the 990-PF forms do not provide adequate data on the composition of foundation assets, as opposed to their aggregate value. Prior to 1982, foundations were required to break their assets down among several different asset classes, such as equities, long-term bonds, etc., but the forms failed to specify that these different classes should be recorded at fair market value rather than book value. Since 1982, the

forms have specified market value but have dropped the separate identification of equities (i.e., common stocks) and fixed-income securities (e.g., bonds). As a consequence, it is impossible to utilize this data source to examine changes in these two very different types of assets or to assess what impact such shifts might have on rate of return.

Return Rate Calculation

In the second place, the Internal Revenue Service does not require foundations to compute a total rate of return or to utilize some of the standard accounting methods for calculating rates of return (e.g., the so-called unit method).[24] As a consequence, we had to develop our own formula for estimating the return rate from the information provided on the form. As we discuss more fully in Chapter Three and Appendix A, this formula approximates a true "unit return" measure, but it is not precise.

Gaps

A third problem with the 990-PF data results from gaps that exist in the data due to loss of forms, failure to file, missing data items, or slowness in converting forms into microfiche records. As a result, of the 1,003 foundations on which we sought financial records for 1979–1984, complete files could be assembled only for 459, and even then only for the 1979–1983 period. The number for which complete files could be assembled for the entire 1979–1984 period was 184. For much of the analysis of investment return and payout performance presented here, therefore, attention focuses on this 1979–1983 sample of 459 foundations. Fortunately, the sample is still broad enough to support inferences both about the different size classes and about the foundation universe as a whole.[25]

Fiscal Year Problems

In addition to these problems with the 990-PF forms, one further problem complicated the analysis: Foundations have different fiscal years. As a consequence, a rate of return for one foundation for 1980 might cover the period beginning in April 1979 and ending in March 1980, while that of another foundation might cover the period beginning in January 1980 and ending in December 1980. This could cause great confusion in the analysis of the data and, in fact, has distorted the findings of a number of previous studies. We dealt with this problem chiefly by computing return rate and payout data for *groups* of years, thus washing out at least part of the distortion introduced by varied fiscal years.

[24] For a discussion of the "unit method," see Chapter 3.

[25] For a discussion of the standard error of the estimates for these samples, see Appendix A.

Time Period

The five-year time period covered in most of the financial data utilized in this report, while still limited, is nevertheless considerably longer than that utilized in most other attempts to examine foundation investment performance. As with any time period, however, this one has peculiarities that must be borne in mind (these are discussed in more detail in Chapter Four).

Most notable, perhaps, were the high interest rates and high inflation rates at the beginning of the period. Reflecting this, the rates of return on short-term government certificates were higher than at any point in their history. In addition, as shown in Table 1.3 below, the annual rate of return on equities, as measured by Standard and Poor's 500 stock index, was also unusually high. Compared to the 8.9-percent inflation-adjusted annual rate of increase during the 1979–1983 period covered here, the Standard and Poor's index was a negative 3.6 percent a year over the prior five-year period 1974–1978. The record over the entire period from 1950 to 1978 was considerably better than this, but still did not equal the 1979–1983 time frame. Even after adjustment for inflation, in other words, the 1979–1983 period was a very favorable one for investment managers, as measured by the standard market indexes. Therefore, some care must be taken in reading into the 1979–1983 results lessons that will apply to all time periods. This is particularly true in view of the stock market plunge that occurred in October 1987, well after the period covered here.

Control Portfolio

Because no time period is ever perfectly "normal," the best way to neutralize the peculiarities of any particular time period in an analysis like this one is to use a control group. By collecting data on the performance of both foundations and a group of comparable organizations (the control group) during *the same time period,* it is possible to hold the effects of the time period constant and factor them out. Because the foundations

and the control-group organizations are operating in the same market conditions, any differences that turn up in their performance can be attributed to their basic character or to the way they have been managed rather than to the impact of the period.

Identification of the appropriate control group against which to compare foundation investment performance was no simple task, however. Pension funds and college endowments were obvious candidates, but each of these operates under a slightly different regulatory regimen and has somewhat different objectives. What is more, given the fiscal year problem mentioned above, we were unable to find a collection of pension funds or college endowments with a fiscal year profile identical to that of our foundation sample.

In view of these difficulties, we simulated the performance of a control portfolio, using existing data sources. We assumed that our control portfolio was a fund composed of 60 percent common stocks, 30 percent high-grade, long-term corporate bonds, and 10 percent short-term paper—roughly the proportions that prior work suggested was applicable to foundations. We then calculated how this composite portfolio would have performed during 1979–1983 if its stock, bond, and short-term paper portions had performed, respectively, like the Standard and Poor's 500, the Salomon Brothers high-grade corporate bond index, and the 90-day Treasury bill rate. The result provided a standard against which to compare the performance of the actual foundations in our sample.[26] As a further control, moreover, we also gathered data on the performance of

[26] We are indebted to Dr. J. Peter Williamson for assistance in the development of these control portfolio rate-of-return figures. The rates of return for each of the three components are taken from the yearbooks published by Ibbotson Associates of Chicago under the title, *Stocks, Bonds, Bills, and Inflation.* Market-rate-of-return data used later in this study employ the same source. These yearbooks are the annual updates of the seminal work by Roger Ibbotson and Rex Sinquefield in developing historical data on capital market rates of return.

Table 1.3
Relative Performance of Standard Stock and Bond Indexes, 1979–1983 versus 1974–1978 and 1950–1983, Adjusted for Inflation

Index	Average Annual Rate of Return (Percent)		
	Study Period, 1979–1983	1974–1978	1950–1978
Standard and Poor's 500 stocks	8.9	−3.6	6.9
Salomon Bros. bond index	−1.6	−1.9	−0.5
Treasury bills	2.7	−0.7	0.1
Composite[a]	5.1	−2.8	4.0

[a] Assumes a portfolio of 60 percent stocks, 30 percent bonds, and 10 percent cash.

certain other funds, such as the NACUBO college and university endowment index, DeMarche Associates, and the Common Fund.[27]

Weighting

In order to blow up our survey and 990-PF samples to represent the entire universe of foundations, weighting systems had to be developed. In particular, we needed two different weighting systems—one to blow up the sample to represent the universe of *foundations* ("foundation-weighted") and one to blow up the sample to represent the universe of foundation *assets* ("asset-weighted"). Generally speaking, the aggregate results reported in the body of this report are the weighted results, with the context determining whether foundation-weighted or asset-weighted results are being reported (see Appendix A for detail on the weighting methods used).

V.
OUTLINE OF THIS REPORT

The purpose of this report is thus to provide a more complete and empirical analysis of the way a cross-section of foundations manages assets and expenditures over time than yet has been available, and to determine the actual rates of return and payout that these foundations have achieved in the recent past. In the process, the report provides a basis for assessing what impact the change in the payout requirement in 1981 may be having on foundation investment and payout practices and performance.

To explore these points, the discussion that follows is organized into five chapters. Chapter Two focuses on the investment management process. It examines the roles and responsibilities of the various entities involved in the three key facets of foundation investment management—objective-setting, strategy-setting, and portfolio management. Chapter Three goes beyond this to look at the actual content of foundation investment objectives and strategies and seeks to determine whether these objectives and strategies changed as a consequence of the 1981 change in the payout requirement. Chapter Four then analyzes the actual *investment performance* of foundations: It reports the rates of return our cross-section of foundations achieved in their investment activities during the period 1979–1983, assesses how well foundations have performed in comparison to our control portfolio, compares performance before and after the change in the payout requirement, and then seeks to explain the variations in performance that emerge. In Chapter Five, attention turns to the *payout* record of foundations and to the relationship between payout rate and return rate before and after the 1981 change in the payout regulations. Chapter Six then pulls these threads of analysis together and evaluates what the data indicate about the management of the foundation investment process, about the impact of the change in the payout requirement enacted in 1981, and about the overall wisdom of the existing regulations on foundation investment and payout behavior. Against this backdrop, it makes some recommendations for the management of the foundation investment process and for the regulations that govern it.

[27] NACUBO, the National Association of College and University Business Officers, gathers data on over 200 college and university endowment funds and reports on investment returns using the "unit method." DeMarche Associates is a private investment reporting firm that includes a number of foundations among its clients. The Common Fund manages a number of investment pools for educational and other nonprofit organizations; the reference to the Common Fund above and in what follows is in fact to a *composite* of 60 percent of the Common Fund equity pool, 30 percent of the Common Fund bond pool, and 10 percent of the Common Fund short-term pool.

Chapter Two

THE INVESTMENT MANAGEMENT PROCESS

By its very nature, investment management is a complicated process because it deals with an uncertain and, indeed, unknowable future. But investment management for foundations involves additional complexities that result from the foundation form and the legal constraints that have grown up around it.

- Legal payout requirements inhibit the growth potential of foundations that may want to do more in the future than they are doing in the present. The more the foundation pays out, the less there is to provide for future grants.

- An excise tax based partly on the size of net capital gains also drains investment capital, a disadvantage not faced by certain other tax-free institutions like pension funds.

- The absence of systematic cash inflows like those enjoyed by pension funds limits the ability of foundations to overcome losses and, possibly, take a longer-term view of the investment process.

- Penalties associated with "jeopardy investments" or requirements to dispose of donor property within specified time periods limit investment discretion.

- Because foundations often received their wealth in the form of stock in the donor's company, they are frequently less diversified than investors whose original assets came in the form of cash rather than securities.[1] Concentration of assets in any one area makes portfolios more volatile than they would be if they were more broadly diversified.

Beyond this, foundation trustees, who have ultimate responsibility for the investment activities of the foundation, often hold their posts because of their interest in the charitable objectives of the institution rather than because of their financial management experience.

How do foundations cope with these challenges? How are foundation investment objectives set, and how are they translated into operation? What roles do foundation trustees, staff, and outside managers play? To what extent do investment decisions drive payout decisions, and to what extent does the opposite occur? What investment objectives are foundations seeking, and how do they balance the competing concerns of controlling risk, ensuring current program levels, and expanding or maintaining the real value of their assets? What change, if any, has the modification of the foundation payout requirement in 1981 made in the way foundations approach their investment activities?

The answers to these questions are important not only for what they can tell us about the main lines of foundation investment management *practice,* but also for the insight they can offer into foundation investment *performance.* The rates of return that foundations achieve, for example, are likely to be affected by the investment objectives they seek and the level of risk they are willing to tolerate. This, in turn, can be affected by

[1] Under the 1969 tax act, foundations were encouraged to diversify their holdings and were penalized if they, or those associated with them (e.g., family members of founders), own more than 20 percent of the voting stock of a corporation.

the extent of involvement of foundation trustees in investment matters and the degree of latitude and support trustees give to investment managers.

Unfortunately, very little is known about these matters. The purpose of this and the following chapter is to fill this gap in knowledge by examining the process and the substance of foundation investment management, drawing chiefly on the results of our mail survey of foundations. The focus of this chapter is the foundation investment management *process,* especially the roles played by different actors in each of three key aspects of the investment process—setting objectives, defining strategy, and day-to-day portfolio management. Chapter Three then examines the *substance* of foundation investment objectives and strategies, including such issues as rate of return goals, investment time frame, and willingness to take risks; it also explores the tangible manifestation of these objectives and strategies as reflected in the *composition of foundation assets.* The goal throughout is to lay the groundwork for the subsequent analysis in Chapter Four of the actual investment and payout performance foundations have achieved.

I.
THE INVESTMENT MANAGEMENT PROCESS: A FRAMEWORK FOR ANALYSIS

Properly handled, the process of investment management involves three interrelated activities: first, the setting of overall *objectives;* second, the formation of a *strategy* for achieving those objectives; and third, the actual day-to-day *management* of the investment portfolio.[2]

Investment objectives establish the "game plan," or the overall guidance for the investment function, specifying the long-term goals of the institution, its tolerance for risk, and the time frame over which performance is evaluated. For example, foundations must decide

whether to seek to maximize the current income they receive on their assets ("income yield"), the long-term growth of the asset base, or some combination of the two ("total return"). These choices can lead to widely different investment approaches. For example, a foundation primarily interested in long-term growth can afford to invest in low-yield stocks with good prospects for appreciation. But these choices are in turn driven by the institution's tolerance for risk. Because returns are positively related to risk in the long run, and because risk means a higher probability of loss in the short run, foundations typically have to choose between short-term stability and long-term growth. Making these broad policy choices is the heart of setting investment objectives.

Important as investment objectives are, however, they are not the whole of the investment process. Objectives must be translated into concrete action terms. To do this, an investment strategy is commonly needed. The role of an investment strategy is to specify the route that is to be taken to achieve the objectives—the types of assets to hold, the proportion of each type to include in the portfolio at any point in time, the division between internal and external management, and so forth.

Finally, strategies must be implemented on a day-to-day basis. Stocks, bonds, real estate, and other assets must be bought and sold. Decisions must be made about particular stocks or bonds to buy or sell at a given point in time. This is the task of "portfolio management."

Inevitably, these three facets of the investment management process flow into each other, and the distinctions cannot be as clear in practice as they are in theory. Yet there is widespread agreement in the literature that all three functions are important and that it is possible, and at times desirable, for different players to be involved in each. Thus, while the principal responsibility for foundation investment activity belongs, like other facets of foundation operations, with the board, the board can fulfill this responsibility by setting overall investment objectives and a broad strategy for carrying these objectives out. Once this is done, the day-to-day management of the portfolio can be left to specialized staff under the board's general supervision.[3]

---

[2] Charles D. Ellis, *Institutional Investing* (Homewood, Il.: Dow, Jones, Irwin, 1971), pp. 10–26; Roger Murray, "Foundation Investments: Problems of Investment Policy," in *Tenth Biennial Conference on Charitable Foundations* (New York University, 1971), pp. 29–35; Burton G. Malkiel and Paul Firstenberg, *Managing Risk in an Uncertain Era* (Princeton: Princeton University Press, 1976); John W. Nason, *Trustees and the Future of Foundations* (New York: Council on Foundations, 1977); William Caryt and Craig B. Bright, *The Law and Lore of Endowment Funds* (New York: The Ford Foundation, 1969); Donald Young and Wilbert Moore, *Trusteeship and the Management of Foundations* (New York: Russell Sage Foundation, 1969); John G. McDonald, "Setting Investment Objectives," Graduate School of Business, Stanford University, Research Paper No. 136 (1973); "Investment Strategy for a Private Foundation," *Foundation News* (January/February 1980), pp. 31–33, 41: Svie Bodie, "Investment Strategy in an Inflationary Environment," National Bureau of Economic Research, Paper No. 701 (1981).

[3] According to *The Handbook on Private Foundations* published by the Council on Foundations in 1981, for example, "Setting investment policy for a foundation is an important function of its board" (p. 67). See also: Ellis, *Institutional Investing,* 1971; Murray, "Foundation Investments"; Nason, *Trustees and the Future of Foundations,* 1977; McDonald, "Setting Investment Objectives," 1973. For a forceful statement of the position that foundation boards should retain an active voice not only in the setting of foundation investment objectives, but also in the day-to-day management of foundation investments, see: David R. Frazer, "Investing Your Foundation's Assets," *Foundation News* (January/February 1985), pp. 46–49.

How well does the investment process of private foundations, as reflected in our survey results, conform to this widely accepted image? To what extent is the board involved in investment management, even if only to set broad objectives?

To answer these questions, we asked our survey respondents to indicate which of a number of potential participants in the investment management process were "extremely important," "somewhat important," "not too important," or "not important at all" in each of the three major facets of investment management. Foundations were permitted to give any of the ratings to more than one participant or sets of participants in the process. For example, a given foundation could indicate that the board of trustees, an investment committee of the board, and an outside manager all play "extremely important" roles in setting investment strategy. Foundations were then grouped into three categories: (1) those that cited the full board of trustees as an "extremely important" actor in a given part of the investment process regardless of what other actor was noted; (2) those that credited not the full board but a portion of the board (e.g., the board chairman or a committee of the board) as being "extremely important"; and (3) those that cited internal or external staff, but not the full board or a portion of the board, as being "extremely important." The following discussion reviews some of the major findings that emerge from this analysis.

II.
PLAYERS AND ROLES IN FOUNDATION INVESTMENT MANAGEMENT

Setting Investment Objectives

More than any other facet of the investment management process, the setting of overall investment objectives or goals is the one where active involvement by foundation boards seems most appropriate and necessary. It is at this stage in the process, after all, that guidance must be provided about the most basic issues of financial management—the level of risk that can be tolerated, the balance between current payout and future asset growth, and so on. What do our data reveal about how these decisions are made in practice?

Limited Role of Foundation Boards

While foundation boards are clearly the dominant players in the setting of investment objectives, as much of the literature in the field would recommend, less than half of the foundations in our sample indicated that the full foundation board was "extremely impor-

tant" in setting investment objectives (see Table 2.1). By contrast, 30 percent indicated that foundation boards were either "not too important" or "not important at all" in setting investment objectives.

While the board as a whole may not have played a crucial role in setting investment objectives for most foundations, subcommittees of the board, or the board chairperson, played such a role for a significant minority of foundations. When we combine these three forms of board involvement in investment objective setting and calculate the proportion of foundations in which *either* the full board, *or* the board chairman, *or* an investment committee of the board is credited with an "extremely important" role, the picture that emerges is somewhat closer to what recommended practice would suggest. In particular, about 64 percent of the foundations indicated that the full board or some portion of it takes an "extremely important" part in setting the foundation's overall investment objectives.

Significantly, however, for over a third of the foundations, not even a portion of the board is taking an active hand in this most basic investment function. And the fact that in most foundations board involvement is either limited overall or delegated to a subgroup or an individual suggests that boards *qua* boards are playing less of a role in setting the broad contours of foundation investment policy than is often recommended.

As Table 2.1 also shows, this pattern of somewhat limited board involvement in foundation objective setting was considerably more evident among the smaller foundations (those with assets under $10 million) than among the larger ones. Also striking is the important role attributed to investment committees of the board in setting investment objectives among the larger foundations. As a result of this, all but 11 percent of the larger foundations credited the board with an "extremely important" role in setting investment objectives through one of the three routes available (full board, board chair, or investment committee). Among smaller foundations, the proportion that failed to credit the board with an extremely important role in objective setting was a much higher 37 percent.

Substantial Role of Outside Managers

Compared to the somewhat less extensive role that foundation boards seem to be playing in the development of investment objectives than might have been expected, bank trust departments and other outside managers are identified as playing a somewhat more substantial role than might have been expected. Almost one in five foundations identified bank trust departments as being "extremely important" in setting investment objectives. This was particularly true among small- and medium-sized foundations, but even among

Table 2.1
Key Participants in Setting Foundation Investment Objectives

	Percent of Foundations Identifying Actor as Playing "Extremely Important" Role			
Actor	All Foundations (N = 22,214[a])	Large Foundations[b] (n = 73)	Medium Foundations[c] (n = 124)	Small Foundations[d] (n = 279)
Full board	45.8	48.6	52.4	45.7
Board chairman or investment committee	18.1	40.3	26.6	17.7
Staff or managers only	36.1	11.1	21.0	36.6
Total	100.0	100.0	100.0	100.0

Source: Salamon/Voytek Foundation Investment Study.

[a] Foundation-weighted results, adjusted for nonrespondents to this question. Throughout this report *N* is used to depict the universe of foundations and *n* to depict sample size.

[b] Foundations with assets of $50 million or more.

[c] Foundations with assets of $10–49.9 million.

[d] Foundations with assets under $10 million.

the largest foundations (those with assets of $50 million or more), 10 percent indicated that bank trust departments played a major role in setting investment objectives (see Appendix B, Table B-1). This suggests that foundation boards are looking to outside managers not just to implement the board's goals, but also to set the goals.[4]

Setting Investment Strategy

Investment objectives merely set the overall goals for foundation investment activity—e.g., whether to go after high current income or growth in the asset base. The foundation still has to develop guidelines about how to implement these goals. Does it invest in all

[4] One caveat that must be borne in mind in interpreting this finding is that for a substantial number of small foundations, banks serve in lieu of boards of directors as the trustees for the foundations. In other words, for these foundations the board and the outside manager are the same.

kinds of securities—stocks, bonds, paper, real estate, tangible assets, assets in foreign countries, and venture capital? Or does it exclude some asset classes or define maximum or minimum holdings for them? The discussion that follows attempts to provide some insight into how foundations answer these questions.

Limited Board Involvement

As reflected in Table 2.2, the proportion of foundations reporting that foundation boards of trustees are "extremely important" in setting investment *strategies* is even more limited than the proportion crediting the board with an extremely important role in setting investment *objectives*. Only 37 percent of the foundations indicated a major role for their full boards of trustees in this function, and the figure did not vary much by foundation size. Among the larger foundations, a considerably higher proportion indicated that the board chairman or an investment committee of the board

Table 2.2
Key Participants in Setting Foundation Investment Strategies

	Percent of Foundations Identifying Actor as Playing "Extremely Important" Role			
Actor	All Foundations[a] (N = 22,210)	Large Foundations[b] (n = 71)	Medium Foundations[b] (n = 124)	Small Foundations[b] (n = 279)
Full board	37.3	35.6	42.8	37.2
Board president or investment committee	23.0	45.2	27.4	22.7
Staff or managers only	39.7	19.2	29.8	40.1
Total	100.0	100.0	100.0	100.0

Source: Salamon/Voytek Foundation Investment Study.

[a] Foundation-weighted result, adjusted for nonrespondents to this question.

[b] See Table 2.1.

Table 2.3
Formal Investment Strategy Guidelines

Type of Guideline	Percent of Foundations with Restriction		
	All Foundations (*N* = 22,214)	Large Foundations[a] (*n* = 197)	Small Foundations[b] (*n* = 279)
Maximum stock holdings	22.0	39.6	21.5
Minimum stock holdings	11.7	20.9	11.5
Maximum fixed-income holdings	17.2	30.6	16.9
Minimum fixed-income holdings	12.4	21.8	12.2

Source: Salamon/Voytek Foundation Investment Study.

[a] Assets of $10 million or more.

[b] Assets of less than $10 million.

played an "extremely important" role in setting investment strategies, but even here the proportion of foundations so reporting was less than half (45 percent). When the three alternative routes of board involvement in investment strategy setting—the full board, the board chairman, or an investment committee of the board—are combined, the results are more encouraging. As shown in Table 2.2, about three-fifths of the foundations credited the board with an extremely important role in setting strategies through at least one of these routes, and among the largest foundations this figure reaches four in five. Yet, this means that for 40 percent of the foundations, board involvement in setting investment strategy is more limited. And even where board involvement is extremely important, it is typically not the full board that acts.

Limited Involvement by Other Actors

Not only did a relatively small proportion of the foundations record a major role for the full foundation board in setting investment strategies, but also an even smaller proportion identified other actors as playing a crucial role in this process. Thus, only between 15 and 20 percent of the foundations identified the executive director or a bank trust department as a key actor in the setting of investment strategies, and only between 10 and 15 percent identified other foundation staff or another outside manager (See Appendix B, Table B-2).

Apparent Inattention to Formal Investment Strategy

Additional data suggest, in fact, that most foundations do not really develop formal investment strategies.

This suggestion finds support in Table 2.3, which records the extent to which foundations report having various types of investment strategy guidelines. As this table shows, less than a quarter of the foundations reported having a formal guideline spelling out the maximum portion of their assets that could be held in common stock, perhaps the most basic kind of guideline that might be expected. Other kinds of restrictions—e.g., on trading in commodities or holdings of real estate—were even less common. Here again, the larger foundations seem to have more formal guidelines than the smaller ones, as might be expected, but even among the large foundations no more than 40 percent indicated any formal guideline, even on the share of assets that could be held in common stock.

Consistent with this, only about 40 percent of the foundations surveyed reported having any formal procedure for reviewing their investment strategies. The conclusion that seems inescapable is that most foundations do not develop formal strategies to guide their investment operations, preferring to operate in a more ad hoc fashion or to leave it to outsiders to make the necessary choices.

Portfolio Management

An ad hoc management style, in which investment strategy is left loose and unspecified, would make sense if those with ultimate responsibility for the foundation—the trustees or senior executives—played a significant role in the third aspect of the investment process, day-to-day portfolio management. Under this scenario, the trustees would develop their strategies and objectives in the course of making specific investment decisions. Guidelines might still be desirable for rational investment decisions, but the formality of the process could be greatly curtailed. To what extent does this alternative image describe the operation of foundation investment management? To what extent do the trustees actually participate in a continuous and systematic manner in the day-to-day management of foundation portfolios?

Table 2.4
Major Participants in Foundation Portfolio Management

	Percent of Foundations Identifying Actor as Playing "Extremely Important" Role			
Actor	All Foundations (*N* = 22,214)	Large Foundations[a] (*n* = 73)	Medium Foundations[a] (*n* = 124)	Small Foundations[a] (*n* = 279)
Full board	22.6	5.6	11.4	23.0
Board chairman or investment committee	18.0	9.7	22.0	18.0
Staff or managers only	59.4	84.7	66.6	59.0
Total	100.0	100.0	100.0	100.0

Source: Salamon/Voytek Foundation Investment Study.

[a] See Table 2.1.

Limited Involvement of Directors

As it turns out, the data generated in our survey lend little credence to this alternative image. On the contrary, as reflected in Table 2.4, our survey indicates that less than a quarter of the foundations credit their full boards of trustees with an "extremely important" role in the day-to-day management of foundation assets. To be sure, the smaller foundations, which less commonly had formal guidelines or strategies, were somewhat more likely than the larger foundations to involve boards in actual investment management although, as mentioned earlier, bank trust departments may function for some small foundations as both the directors of the foundation and the managers of the assets. But only about a quarter of the smaller foundations reported that their full boards of trustees play an extremely important part at this stage of the process. What is more, this conclusion does not change much when we relax the definition of board involvement to include not only involvement of the full board but also of the board chairman or an investment committee. In only 41 percent of the foundations, as shown in Table

2.4, did boards play an important role in portfolio management through any of these routes.

Extensive Reliance on Outside Managers

At least as important as board members and senior staff in the actual business of managing foundation portfolios are bank trust departments and other outside managers. In fact, as reflected in Table 2.5, just over 40 percent of the foundations vest day-to-day portfolio management in outside managers of some sort, and most of these rely on such managers to handle two-thirds or more of the foundation's assets.

This reliance on outside managers is particularly evident among the largest foundations, those with assets exceeding $50 million. Close to 90 percent of these foundations rely on outsiders to handle at least some of their assets, and nearly three out of four use outside managers for two-thirds or more of their assets. Such delegation may make sense for these foundations since they are the ones where the trustees most often define explicit investment objectives and set investment strategy. Here, in other words, guidelines exist to give

Table 2.5
Share of Foundation Assets Managed Outside, by Size of Foundation

	Percent of Foundations Recording Given Share of Assets Managed Outside			
Share of Assets Managed Externally (Percent)	All Foundations (*N* = 22,214)	Large Foundations[a] (*n* = 73)	Medium Foundations[b] (*n* = 124)	Small Foundations[c] (*n* = 279)
0	58.2	12.5	24.6	59.3
1–33	2.0	8.3	5.9	1.8
33.1–67	2.0	6.9	8.5	1.8
Over 67	37.8	72.2	61.0	37.0
Total	100.0	100.0	100.0	100.0

Source: Salamon/Voytek Foundation Investment Study.

[a] Assets of $50 million or more.

[b] Assets of $10–49.9 million.

[c] Assets of less than $10 million.

Table 2.6
Pattern of Foundation Reliance on Outside Managers

Number of Outside Managers	Percent of Foundations Using Given Number of Managers			
	All Foundations (*N* = 22,210)	Large Foundations[a] (*n* = 73)	Medium Foundations[a] (*n* = 124)	Small Foundations[a] (*n* = 279)
None	58.2	12.7	25.4	59.1
1	38.2	29.6	50.0	38.0
2–3	3.4	14.1	20.2	2.9
4 or more	0.2	43.7	4.4	—
Total	100.0	100.0	100.0	100.0

Source: Salamon/Voytek Foundation Investment Study.

[a] See Table 2.5.

direction to outside managers in carrying out their day-to-day activities.

By contrast, only 40 percent of the small foundations make any use of outside managers, and these are the foundations that are less likely to provide formal guidelines for outsiders to follow. The preference for inside management may simply be the result of the modest size of the foundation and the larger role of the trustees relative to professionals in the running of the whole operation. In addition, as noted earlier, for many of these foundations, the roles of trustee and outside manager may be merged in the form of a bank trust department that plays both roles. Finally, the high cost of retaining outside managers may deter foundations with small portfolios from taking this route.[5]

Use of Multiple Managers

In addition to placing a substantial share of their assets in the hands of outside managers, large foundations also tend to utilize more than just one manager. As noted in Table 2.6, 58 percent of the large foundations use two or more managers, and 44 percent use four or more. By contrast, smaller foundations that make use of outside managers tend to rely on only one manager, most probably a bank trust department.

Discretion of Outside Managers

One final point about outside managers has to do with the degree of discretion granted to them by their foundation clients. Generally speaking, foundations vest a substantial degree of discretion in outside managers. Over half of the foundations that utilize outside managers indicate that they give their managers "complete autonomy and discretion" in day-to-day portfolio management, and another 30 percent indicate that their managers enjoy a "high degree of autonomy and discretion." Clearly, outside managers, where they are used, are significant shapers of foundation investment behavior.

[5] Small foundations may use mutual funds to overcome this problem. In a sense, the fund managers become the foundation's outside managers.

III.
SUMMARY AND CONCLUSIONS

One conclusion that begins to emerge from the data presented here is that the foundation world seems to contain two very different modes of investment management. One of these operates the way conventional investment wisdom recommends, with foundation trustees setting objectives, working with staff to translate these objectives into concrete strategies, and then leaving to outside managers the task of implementing the objectives and strategies through day-to-day portfolio management. The other operates in a much more ad hoc and undirected way, with foundation trustees and staff essentially leaving to external or internal managers much of the responsibility for foundation investment policy making, or with bank trust departments operating as policy setters and managers all rolled into one.

Although we cannot determine with precision what share of the foundations fall into each of these modes and what share fall in the middle, the preponderance of the evidence reported here suggests that the ad hoc route is by far the more common. Most foundations do not appear to involve their full boards in the setting of objectives and the specification of strategies. Therefore, the investment function is allowed to float, picking up its cues from the professional managers inside and out. This is understandable since the vast majority of these foundations (82 percent) report having no staff, and many have no separate boards of directors as such, vesting this function instead in bank-appointed trustees. The one countervailing consideration is that the more structured approach appears to be much more common among the larger foundations, so that the larger part of foundation *assets* may be handled in the recommended way even though most of the foundations manage the function far less directly and formally.

The important question, however, is not simply what the process is for making foundation investment decisions, but what the *content* of those decisions is. It is to this question that we turn in the next chapter.

Chapter Three

FOUNDATION INVESTMENT GOALS AND STRATEGIES

If, as Chapter two suggests, the majority of foundations do not seem to manage their investment function with the degree of formality or board involvement that many investment specialists recommend, this does not necessarily mean that foundations are pursuing the wrong goals or using the wrong management style. Those in actual control of foundation investments may still be pursuing a reasonable approach even though the full boards of the foundations are not active players in the process.

To evaluate foundation investment management fairly, therefore, it is necessary to look beyond the investment management *process* at the actual *content* of foundation investment objectives and goals. It is the purpose of this chapter to examine these goals and strategies and to assess what changes have occurred as a consequence of the modification of the foundation payout requirement in 1981.

Four key facets of investment strategy are of particular interest: the rate-of-return goal, the degree of risk tolerance, the time frame over which performance is judged, and, finally, the nature of the management style used. These four elements are obviously interrelated. The central concept is the one to which we referred earlier: the linkage between risk and expected return. A foundation aiming at a high rate of return must normally accept a higher level of risk than one content with a lower rate of return. This in turn can affect the

kinds of assets the foundation acquires. No one expects to earn as much from an insured savings account as they might earn from wildcatting for oil. On the other hand, no one expects that the probability of loss is as great on the savings account as it would be in the search for oil. Markets therefore tend to price assets so that the expected rewards bear some kind of systematic relationship to their perceived risks of loss: Investors pay less for a dollar's expected return on wildcatting than they would pay for a dollar's expected return from an insured savings account.

This trade-off is sometimes a matter of choice. Some investors are more risk averse than others, and some are in a better financial position to absorb losses than others. In other cases, inadvertence rather than conscious choice governs the process. Finally, the trade-off is sometimes a consequence of necessity and grows out of a foundation's grant-making strategy. Foundations that want to give away more money now than later cannot take a long-run view of the capital markets. They cannot survive if they have to liquidate assets in depressed markets. Similarly, foundations with a firm commitment to fund large projects in the future in effect carry a fixed liability and cannot afford the risks of assets with variable or uncertain incomes.

The measurement of some of these dimensions of investment activity can be quite complex, however. For example, elaborate approaches have emerged in recent years to characterize and measure risk.[1] But most of these would have been unfamiliar to the respondents to our survey. Fortunately, however, it was possible to develop a more intuitive approach and terminology that captured most of the basic concepts and relationships.

In the balance of this chapter, we first examine how foundations define their investment objectives and strategies and then look at how these decisions show up in foundation asset composition.

I.
THE CONTENT OF FOUNDATION INVESTMENT OBJECTIVES

Rate-of-Return Goals

At the core of the development of an investment approach is the selection of a rate-of-return goal. Inevitably, this involves a trade-off between current income and future income. Assets that yield high income in the short run rarely have great prospects for asset growth over the long run. If they did, investors would bid up their price, thus reducing their current yield and growth potential. In practice, therefore, foundations must explicitly or implicitly choose to maximize either their current income or the long-term value of their assets, or some combination of the two. In other words, they must decide whether their rate-of-return goal is "income maximization," "capital appreciation," or "total return."

[1] Perhaps the most well developed of these theories is the "capital-asset pricing theory" or "modern portfolio theory." According to this theory, there are two basic types of risk associated with any investment: first, "systematic risk," which is the degree of sensitivity of an investment to overall market movements; and second, "unsystematic risk," which is the variability or volatility of a particular investment associated with factors peculiar to that investment rather than with overall market trends. Modern portfolio theory holds that "unsystematic risk" can be reduced through diversification, but that "systematic risk" cannot. Investors must therefore choose the level of "systematic risk" they can tolerate. To help them, a measure of the systematic risk associated with a security—known as *Beta*—was developed. Beta essentially measures the historic relationship between the movements of a particular stock (or portfolio) and the movements of the market as a whole. A security with a Beta of 2 has thus on average moved up or down twice as much as the market averages—e.g., going up by 10 percent when the market rises 5 percent. For an excellent discussion of Beta and modern portfolio theory, see: Burton G. Malkiel, *A Random Walk Down Wall Street*, 4th Ed. (New York: W. W. Norton and Co., 1985), pp. 185–209. For more technical treatment, see: Franco Modigliani and Gerald Pogue, "An Introduction to Risk and Return, *Financial Analysts Journal* I (March–April 1974), II (May–June 1974); Robert A. Levy, "Beta Coefficients as Predictors of Return," *Financial Analysts Journal* (January–February 1974); Harry Markowitz, *Portfolio Selection: Efficient Diversification of Investments* (New York: John Wiley and Sons, 1959).

One of the major criticisms of the Tax Reform Act of 1969 was that it discouraged a balanced "total-return" approach because it required foundations to pay out the full amount of their investment income each year. As a result, foundations were "penalized" for investing in high-yielding bonds.[2]

Total-return Objective Predominates

Table 3.1 shows that total return—a combination of appreciation and investment income—is by far the most popular investment objective. At the same time, less than half of the foundations in our survey reported pursuing a total-return objective as of 1984, when our survey was completed.

A striking disparity exists between the investment objectives of large and small foundations, however. Over 80 percent of the larger foundations identified the maximization of total return as their investment objective. By contrast, only 45 percent of the small foundations identified the maximization of total return as their principal investment objective. Rather, larger proportions of the small foundations focused on income maximization, maintaining asset value, and achieving program spending levels. Only 2.8 percent of the large foundations focus on the spending goal, and none of them reported focusing principally on maximizing income.

This disparity seems to suggest a greater degree of investment sophistication on the part of the larger foundations. As such, it is consistent with findings reported above about the lack of coherence and structure in the investment processes of many smaller foundations. Many small foundations may in fact not be pursuing a coherent investment strategy at all but rather trying to minimize their risk of loss by placing their assets in highly secure bonds or other relatively conservative investments and accepting whatever return rate and payout level results.

An alternative explanation may be that small foundations are more interested than the large foundations in spending money in the short run than over the long run, perhaps because the dominant influence in a small foundation is more likely to be the original donor interested in spending "his" or "her" money, in contrast to a large foundation run by a professional staff. Either way, this would lead us to expect a more conservative investment approach on the part of the smaller foundations.

[2] Williamson, "Inflation and the Foundation Payout Rate," 1981, pp. 20–24. Some foundations may have tried to offset this disadvantage by buying low-coupon bonds at a discount and with assured appreciation to maturity, or by buying stocks where the underlying corporation reinvested the earnings of the company instead of paying out the earnings in dividends (Peter Bernstein, personal communication, January 1988).

Table 3.1
Investment Objective, by Size of Foundation, 1984

Objective	Percent of Foundations with Indicated Objective			
	All Foundations (N=22,135)	Large Foundations[a] (n=72)	Medium Foundations[a] (n=124)	Small Foundations[a] (n=278)
Total return	46.0	80.6	64.5	45.3
Income maximization	25.0	—	11.3	25.5
Capital appreciation	5.4	1.4	8.1	5.4
Maintain asset value	11.1	12.5	8.1	11.2
Achieve spending goal	6.7	2.8	3.2	6.8
Other	5.8	2.8	4.8	5.8
Total	100.0	100.0	100.0	100.0

Source: Salamon/Voytek Foundation Investment Study.

[a] See Table 2.5.

Increased Focus on Total Return Since 1981

Whatever the explanation of current patterns, a significant shift in the definition of rate-of-return goal occurred among foundations after 1981, when the payout requirement changed. While a total-return objective still characterized less than half of all foundations as of 1984, this was an increase over the proportion pursuing this objective prior to 1982. Based on our survey, only 35 percent of all foundations pursued a total-return goal prior to 1982, compared to 46 percent after 1982. This is consistent with one of the arguments put forward in support of the 1981 change: that eliminating the requirement that foundations must pay out all of their investment income would free foundations to maximize their total return without worrying about whether the return took the form of appreciation or income.

How important the change in the payout requirement was to this shift toward a total-return objective among foundations is difficult to say, however. Of those who shifted their investment objective since 1981, only about 13 percent indicated that the change in the payout requirement was "extremely important" in leading to the change, while another 26 percent indicated it was "somewhat important." By contrast, about 50 percent reported that the payout requirement change was "not too important" or "not important at all." Therefore, although there is evidence here that the change in the payout requirement may have had some of the effect that was intended, the evidence is far from overwhelming. It may be that the exciting rates of capital appreciation in both the stock market and the bond market after 1981 just changed investor perceptions of what investing was all about.

Relatively Conservative Rate-of-Return Goals

Quite apart from the type of rate-of-return goal foundations are seeking is the numerical value of the return rate being sought. We asked foundations to identify the actual return, including both earnings and appreciation, that they were seeking over the next three years, before adjusting for inflation. The responses, recorded in Table 3.2, reveal a relatively conservative set of investment expectations on the part of foundations as of the mid-1980s.

At a time when the Standard and Poor's 500 had generated a rate of return of 16.5 percent over the prior three years, and the Salomon Brothers long-term corporate bond index had returned 20.6 percent, 92 percent of the foundations in our sample reported a three-year rate-of-return goal of less than 13 percent. The rate-of-return goals among the larger foundations were more aggressive, but even here well over a third reported total three-year rate-of-return goals of less than 10 percent, and only a quarter reported rate-of-return goals of 13 percent or more.

Evidently, the majority of foundation investment managers were skeptical that recent market trends would continue and therefore hedged their expectations.

Table 3.2
Foundation Rate-of-Return Goals

Three-year Rate of Return (Percent)	Percent of Foundations Seeking Given Total Rate-of-Return Goal		
	All Foundations (N=21,195)	Larger Foundations[a] (n=192)	Smaller Foundations[b] (n=266)
3–5	2.3	3.1	2.3
6–9	48.1	33.7	48.5
10–12	41.6	36.8	41.7
13 or more	8.1	25.9	7.5
Total	100.0	100.0	100.0

Source: Salamon/Voytek Foundation Investment Study.

[a] Assets of $10 million or more.

[b] Assets under $10 million.

Risk Tolerance

Rate-of-return goals are in turn related to risk toler-ance, the level of risk foundations are willing to accept. As we pointed out above, the higher the rate-of-return goal, the greater the risk that must be taken. By the same token, the lower the risk one is willing to tolerate, the lower the rate of return one is likely to achieve. Risk tolerance is also related to the time frame over which investments are managed, however, and the de-gree of variability in return that is acceptable.

Although these generalizations are familiar, measur-ing risk is a subtle and tricky matter. Risk is often in the eye of the beholder rather than being some objec-tive quantity. At the same time, efforts to define and measure risk in a precise quantitative fashion involve statistical complexities that lie far beyond the under-standing of most investors.

Investment theory tells us that the most effective measure of risk relates to the variability in the returns on a particular asset in question. All investors like up-ward variability, which is what makes this concept less than totally acceptable, but variability does serve as a useful proxy for risk because of its close relationship to uncertainty. Our confidence in where we are going to end up is a lot higher on a smooth road than it is on one with many turns and bumps. However, the road with twists and turns may be the only one that leads up the mountain.

We attempted to capture this idea in language that our respondents would be likely to understand. Accord-ingly, we asked them to express the level of risk they were prepared to tolerate by comparing it to the risk associated with some of the standard bond and stock indexes. In particular, a low level of risk was defined as that associated with the popular bond market averages, and a high level as that in excess of popular stock indexes.

The result, reported in Table 3.3, shows that the risk tolerance of the foundations represented in our sample appears to be fairly limited. Seventy percent of the re-porting foundations expressed risk tolerances at or be-low those associated with popular bond market averages. Only about a quarter of them were comfort-able with the risks associated with owning the com-mon stocks listed on the popular stock averages (such as the Dow Jones).

These results are consistent with the fairly low rate-of-return goals identified by these foundations. In other words, the majority of foundations appeared willing to accept lower rates of return as the price of avoiding greater risks.

Table 3.3 also shows that risk tolerance varied sharply with the size of the foundation, however. Only about one-third of the largest foundations classified themselves in the very low or moderately low risk-tolerance groups, while almost 75 percent of the small foundations reported low risk tolerance. This, too, is consistent with earlier findings. In fact, it provides fur-ther evidence of a "bimodal" distribution of founda-tions, with one group operating in a more aggressive and activist mode characterized by active board in-volvement in the setting of priorities, an aggressive rate-of-return goal, and a relatively high risk tolerance while the other group is operating in a more conservative and less aggressive fashion. While the fit is far from perfect, moreover, the first group tends to contain a dispropor-tionate number of the largest foundations and the latter group a disproportionate number of the smallest foun-dations. Perhaps their smaller size leaves the smaller

Table 3.3
Risk Tolerance of Foundations

Risk Level	Percent of Foundations with Given Risk Tolerance			
	All Foundations (*N*=21,188)	Large Foundations[e] (*n*=68)	Medium Foundations[e] (*n*=122)	Small Foundations[e] (*n*=266)
Very low[a]	31.7	5.9	13.1	32.3
Moderately low[b]	40.6	29.4	45.1	40.6
Moderately high[c]	24.2	57.4	37.7	23.7
Very high[d]	3.4	7.4	4.1	3.4
Total	100.0	100.0	100.0	100.0

Source: Salamon/Voytek Foundation Investment Study.

[a] Less than popular bond market averages.

[b] About equal to popular bond market averages.

[c] About equal to popular stock market averages.

[d] Greater than popular stock market averages.

[e] See Table 2.5.

Table 3.4
Foundation Investment Time Horizon

Time Horizon	Type of Foundation (Percent)			
	All Foundations (*N*=22,287)	Large Foundations[a] (*n*=73)	Medium Foundations[a] (*n*=124)	Small Foundations[a] (*n*=274)
One year	46.3	27.4	34.7	46.7
Three years	14.4	17.8	18.6	14.2
Five years	15.4	41.1	27.4	15.0
Other or don't know	14.4	13.7	19.4	24.0
Total	100.0	100.0	100.0	100.0

Source: Salamon/Voytek Foundation Investment Study.
[a] See Table 2.5.

foundations feeling they have less capacity to absorb losses and therefore inclines them against taking the risk of incurring such losses. In the process, however, it also necessarily lowers their expectation of gain.

Time Horizon

A third key element in investment decision making is the time horizon over which investments are managed. Although security markets are volatile, they do exhibit clearly defined long-term trends. Therefore, the longer the time horizon over which a manager can operate, the more that manager can assume short-term risks in the expectation of earning higher long-run rewards. This is so because there will be time to recover from temporary setbacks and, thus, a greater willingness to enter markets where the prospects for above-average gains are more favorable, but where the chances of short-term setbacks are also more likely.

The investment time horizon is also relevant to the grants strategy a foundation pursues. We should expect foundations that are interested in giving money away sooner rather than later to place more emphasis on current income and less emphasis on growth than would be the case for foundations interested in giving away more money in the future than in the present.

Relatively Limited Time Horizon

In view of the data presented earlier, it should come as no surprise to find that the largest concentration of foundations in our sample reported investment time horizons of only one year. Table 3.4 shows that 46 percent of the respondents reported that horizon, while only 15 percent reported an investment time horizon of five years or more.

As we would also expect, the designated time horizon is much shorter among small foundations than among large ones. Among the large foundations, 41 percent reported a time horizon of five years or more compared to 15 percent for the small foundations. These findings are consistent with the evidence reported

earlier that large foundations have risk tolerances and investment policies that are more aggressive than the approaches taken by small foundations. The larger foundations can afford to take more risks because they are managing their investments over a longer time span and therefore can afford more short-term volatility.[3]

Table 3.4 also shows that more than 10 percent of the foundations failed to specify their investment time horizon, with nearly one out of four of the small foundations falling into this group. This suggests that many foundations may be giving inadequate attention to a critical variable in the management of their assets.

Slight Shift Toward Longer Time Horizon Since 1981

The proportion of foundations with a one-year time horizon barely changed between 1981, when the payout requirement was changed, and 1984. This is contrary to what we had expected to find, since in theory at least, the liberalization of the payout requirement freed foundation managers to pursue a longer-term rate-of-return goal by reinvesting part of the foundation's income instead of paying out 100 percent of it. Perhaps not enough time had elapsed to allow foundation investment managers to adapt to this change. Among the foundations represented in our survey, at any rate, only about a quarter of those indicating a change in their time horizon credited the change in the payout requirement with the shift. One reason for this may be that the retention of a one-year time frame for computation of the required payout in the revised law may be keeping the time horizon of foundation investment managers more constrained than it might otherwise be, despite the other liberalization achieved in 1981.[4]

[3] As we demonstrate later, these differences are also consistent with practices with respect to grant payout policies.

[4] Foundations have until the end of the following tax year to make up the 5-percent payout for any given year. They can also carry over excess distributions for up to five years and use them to meet their payout in any year that they fall short. See footnote 9 in Chapter One for a discussion of the carry-over provision.

Investment Management Style

The various aspects of foundation investment management discussed above—rate-of-return goal, risk tolerance, and time horizon—are not, as we have emphasized, independent of one another. Rather, they are integral parts of an investment approach and as such are interrelated. Therefore, in addition to examining the various aspects of investment approach separately, our survey also attempted to develop an overall measure of investment management style.

For this purpose, survey respondents were asked whether the investment management style of their foundation could best be characterized as "passive," "moderately active," or "very active." As used here, these terms were intended to relate not to the composition of the investment portfolio, but to the degree of energy and activity devoted to investment management. Thus, it is quite possible to have an actively managed portfolio of fixed-income securities.

In particular, *passive management* was defined as involving little turnover in the investment portfolio, review by foundation officials or board members of investment performance on only a quarterly basis or less, and little use of outside data and analysis on the part of foundation officials; *moderately active management* was defined as involving "limited turnover, review of performance at most on a monthly or quarterly basis, and limited use of external data or analysis"; and *very active management* was defined as involving "considerable turnover and activity, weekly or monthly review of performance, constant reevaluation of investment strategies in light of market conditions, and heavy use of outside data and analysis."

The responses to this question, recorded in Table 3.5, provide additional insight into the way the foundation investment process is handled. What is more, the responses are consistent with the other findings presented above.

Limited Proportion of Foundations with Active Management Style

The first conclusion that emerges from Table 3.5 is the relatively small proportion of foundations that reports a very active management style. In particular, only 12 percent of all foundations represented in the sample identified their management style as very active. By contrast, 50 percent identified their style as "passive."

Active Style More Common Among Larger Foundations

A dramatic contrast exists, however, between the management styles of the large and small foundations. Almost all (93 percent) of the large foundations consider themselves either moderately active or very active investment managers. By contrast, less than half (48 percent) of the small foundations put themselves in these two categories.

Limited Use of Investment Reporting Services

Further evidence of the generally passive management style used by the bulk of foundations is also evident in foundation responses to a question about outside investment performance reporting services. Despite shortcomings in coverage, such services offer foundation managers a benchmark against which to compare the performances of their own foundations and, therefore, are potentially important management tools. Nevertheless, only about one in ten of the foundations we surveyed indicated that they subscribe to such services.

One reason for this disinterest in performance reporting services may be evident in foundation responses to another question that asked how important comparisons to other foundations are in their own decision-making. Only about 12 percent of the foundations indicated that comparisons with other foundations or in-

Table 3.5
Foundation Investment Style

Style	Type of Foundation (Percent)			
	All Foundations ($N=21,188$)	Large Foundations[a] ($n=68$)	Medium Foundations[a] ($n=122$)	Small Foundations[a] ($n=266$)
Passive	50.3	6.9	14.6	51.5
Moderately active	37.7	51.4	61.8	37.0
Very active	12.0	41.7	23.6	11.5
Total	100.0	100.0	100.0	100.0

Source: Salamon/Voytek Foundation Investment Study.

[a] See Table 2.5.

vestment pools were important in their own investment decision making, while close to 80 percent indicated they were "not too important" or "not important at all."

Summary: Investment Approach Score

The data on the content of foundation investment strategy reported here thus provide further confirmation of the existence of two very different groups of foundations as far as investment practices are concerned. The first group, which contains most of the foundations, manages its investments in a relatively conservative fashion, with little board or staff involvement, limited tolerance for risk, narrow time horizon, and relatively modest rate-of-return expectations. The other group, which contains most of the larger foundations, manages its assets in a more aggressive fashion, with active board involvement, more clearly defined investment objectives, a higher level of risk tolerance, a longer time perspective, and higher rate-of-return goals.

To gauge this notion more systematically, we developed a composite "investment management score" or index combining the responses foundations provided to the six different dimensions of foundation investment management discussed above—type of investment objective, rate-of-return goal, risk tolerance, time frame, existence of formal procedures for reviewing investment strategies, and management style. We assigned higher points to the more active and aggressive responses. Thus, a foundation reporting a three-year rate-of-return goal in excess of 13 percent received two points, one with a rate of return goal between 10 and 12 percent received one point, and one reporting a rate of return goal below 9 percent received zero points. Similar index scores were assigned for the other dimensions examined in the survey. We then summed these points for each foundation and separated the foundations into three groups based on their cumulative score, with the high-

est scores denoting the foundations with the more activist and aggressive investment approach.[5]

As Table 3.6 makes clear, close to two-thirds of the foundations end up with index scores that are low, indicating a relatively inactive investment approach, low risk tolerance, focus on objectives other than total return, limited time horizon, and so forth. At the other extreme, only about 5 percent of all foundations registered index scores that are high. The remaining 30 percent fell into a medium category.

As we would expect from the earlier discussion, these patterns vary widely by foundation size class. Only 12 percent of the large foundations had low index scores while almost half fell in the higher category. On the other hand, two-thirds of the small foundations are in the low category and only 4 percent in the highest category.

Taken together, these results lend credence to the notion that there are several different worlds of foundation investment management in operation. One of these worlds—characterized by a generally inactive and risk-averse style—is populated by most of the small foundations, a third of the medium-sized foundations, and one in ten of the large foundations. A second—characterized by a much more aggressive and activist style—is the domain of about half of the large foundations, one in five of the medium-sized foundations, and only a small share of the smaller ones. A sizable middle ground is occupied by between 40 and 50 percent of the large- and medium-sized foundations and about 30 percent of the small ones.

These findings underline the dangers involved in making broad generalizations about foundation invest-

[5] Altogether, there were nine possible points. Foundations with zero to three were ranked in the low group, those with four to five in the middle group, and those with six or more in the high group. The method used here is a standard index construction approach.

Table 3.6
Foundation Investment Approach Score

	Percent of Foundations			
Investment Approach Index	All Foundations (N=21,188)	Large Foundations[a] (n=68)	Medium Foundations[a] (n=122)	Small Foundations[a] (n=266)
Low	65.4	12.3	33.1	66.5
Medium	30.1	41.5	47.9	29.6
High	4.5	46.2	19.0	3.9
Total	100.0	100.0	100.0	100.0

Source: Salamon/Voytek Foundation Investment Study.

[a] See Table 2.5

ment management. At a minimum, generalizations must be qualified by a recognition of the two or three distinctive types of investment styles that characterize the foundation universe.

In addition, these findings also make it clear that while size is a major factor in explaining differences in foundation investment style, it is by no means the only such factor. Not all large foundations scored high on the investment management score, and not all small foundations scored low. Indeed, because of the disproportionate *number* of small foundations, there are actually a greater number of small foundations in the high category than there are large foundations.[6] While the fit between management approach and size is statistically significant, therefore, it is by no means complete.

II.
ASSET COMPOSITION

Ultimately, investment objectives and strategies find tangible manifestation in decisions about asset composition (the shares of assets to invest in stocks, bonds, land, commodities futures, etc.) and, to a lesser extent, about the particular choices made within each of these categories. Therefore, it is useful to go beyond the analysis of avowed investment goals and policies to examine the actual composition of foundation investment portfolios.

As we have noted, the data available for such an examination are limited. Prior to 1982, foundations were asked regularly to report on their 990-PF forms the share of the investment portfolios they held in each of seven classes of assets: cash, savings and temporary investments, government obligations, corporate bonds, corporate stocks, land and depreciable assets, and others. However, the forms did not specify that asset values were to be recorded at "market value," i.e., at their current value on the market as opposed to their purchase price or "book value." Thus, there is reason to doubt the accuracy of the data as a measure of actual portfolio composition. Since 1982, foundations have been asked specifically to report the market value of the different components of their investment portfolios. However, the categories were reduced significantly, and the three main categories of assets—corporate stocks, corporate bonds, and government bonds—were merged into one on the form. As a consequence, we cannot determine from the 990-PF forms since 1981 what share of foundation assets are held in fixed-in-

come securities and what share in equities or stock holdings.[7]

Although a precise picture of foundation asset composition over the period we are examining is therefore unavailable, we can shed at least some light on this question based on the data that are available. These data are recorded in Table 3.7.

Overall Pattern

Generally speaking, Table 3.7 confirms the commonly accepted impression that 60 percent of foundation assets are held in stocks, 30 percent in fixed income securities (corporate and government bonds), and the remainder in cash.[8] As of 1979, about 61 percent of foundation assets were in common stocks, about 29 percent in bonds, 11 percent in government bonds, and 18 percent in corporate bonds.

Table 3.7
Composition of Foundation Investment Portfolios, 1979 versus 1983

Asset Type	Percent of Assets	
	1979[a]	1983
Government bonds	11.0	N.A.[b]
Corporate bonds	17.6	N.A.
Corporate stock	60.7	N.A.
Subtotal, securities	89.3	86.3
Cash, savings, temporary investments	2.8	6.2
Land, depreciable assets, mortages	1.8	4.3
Other	6.1	3.2
Total	100.0	100.0

Source: Salamon/Voytek Foundation Investment Study.

[a] Some assets may be recorded at book value.

[b] N.A., not available.

Furthermore, this asset mix does not seem to have changed much over time. Although the combined share of assets in "securities" (including both stocks and bonds) dipped slightly between 1980 and 1981, this may reflect in practice the revaluation of the land, buildings, and depreciable assets from book to market value on the reporting forms rather than any change in investment behavior. Table 3.7 shows that such depreciable assets grew from 1.8 percent of asset value in 1979 to 4.3 percent in 1983.

---

[6] The 46 percent of large foundations scoring "high" on our "investment approach index" thus represent 78 foundations. By comparison, the 4 percent of small foundations scoring high on this index represent 842 foundations.

[7] An attempt was made to overcome this problem by gathering more detailed composition data through our survey, but this proved unsuccessful because of the spotty responses from foundations and because foundations group their assets differently in their own internal reporting.

[8] See, e.g., Williamson, "Investment Expectations and the Foundation Payout Rate, 1976."

Table 3.8
Share of Foundation Assets in Common Stock, 1980, by Foundation Size

	Percent of Foundations			
Share of Assets in Stock (Percent)	All Foundations ($N=22,294$)	Large Foundations[a] ($n=88$)	Medium Foundations[a] ($n=252$)	Small Foundations[a] ($n=119$)
60 or more	13.0	60.0	33.0	12.1
40–59.9	8.6	22.5	28.7	8.1
Less than 40	78.4	17.5	37.4	79.9
Total	100.0	100.0	100.0	100.0

Source: 990-PF Data, Salamon/Voytek Foundation Investment Study.
[a] See Table 2.5.

Variation by Size of Foundation

Perhaps the most important finding that emerges from the data on asset composition is the great variation that seems to exist among foundations in different size classes. The 60 percent of total foundation assets invested in stocks as of 1980 turn out to reflect the investment behavior of only a relative handful of foundations. As Table 3.8 shows, only one out of eight of all foundations were 60 percent or more invested in stocks as of 1980, but three out of five of the large foundations had 60 percent or more of their assets in stock. On the other hand, four out of five of all foundations— but only one out of five of the large foundations—had less than 40 percent of their assets invested in stocks. In short, the conventional wisdom about how foundations invest their assets may apply to most foundation assets, but it does not accurately characterize most foundations. The bulk of the small foundations, which comprise most foundations, are more heavily invested in fixed-income securities.

Even after taking account of the "book value" problem, these data thus provide further confirmation of the conclusion reached above, i.e., that the foundation investment world is far from homogeneous. There are at least two rather different patterns in evidence. While most of the large foundations are more aggressive and less risk averse in their investment behavior, most of the smaller foundations are pursuing more conservative investment strategies focused on income yield rather than total return, and investing a significantly larger share of their resources in fixed-income securities.

Impact of Change in Payout Requirement

One final issue worth exploring has to do with the impact on asset composition of the 1981 change in the payout requirement. One of the major criticisms of this requirement, after all, was that it deterred investment in high-yield, fixed-income securities because it required that all of the income from such investments be paid out in "qualifying distributions." To what extent did

the change in the payout requirement in 1981 lead to a change in the composition of investment assets?

Unfortunately, the change in the Form 990-PF reporting requirements discussed earlier makes it impossible to get a firm handle on the answer to this question. We did, however, ask our survey respondents whether they had made changes in their investment strategy since 1981 and, if so, what role the change in the payout requirement had made in their decisions.

The results, reported in Table 3.9, give some support to the argument advanced in favor of the payout change. In particular, one in five of the foundations surveyed indicated that they had decided to invest more heavily in fixed-income securities since 1981, and another 9 percent reported greater reliance on high-yield stocks. Of those reporting heavier investment in fixed-income securities, moreover, almost 65 percent credited the change in the payout requirement as an important factor in the shift. While there is also evidence of movement in the opposite direction, and although there were also valid investment considerations to explain the shift toward bonds after 1981, the data nevertheless provide

Table 3.9
Reported Changes in Foundation Asset Mix Since 1982

Nature of Change	Percent of Foundations
Increased reliance on fixed-income assets	20.9
More emphasis on short-term investments	13.8
More reliance on stocks	12.0
More reliance on high-yield, low-growth stock	9.2
More reliance on commodities, options	0.4
More concern about social objectives	4.7

Source: Salamon/Voytek Foundation Investment Study.

some support for the argument that the elimination of the payout requirement freed foundations to diversify their investments in the ways that the advocates of the change hoped.

III.
CONCLUSION

Based on the data reviewed so far, *Institutional Investor*'s 1968 characterization of the world of foundation investment management as "a place untouched by the revolution in money management," whatever its validity at the time, seems much less valid today. From all indications, a sizable share of the larger foundations, a proportion of the medium-sized foundations, and substantial numbers—if not proportions—of the small foundations seem to have adopted an approach to investment management that is in reasonable accord with the more sophisticated thinking in the field, involving trustees in setting specified investment objectives, focusing on total return, pursuing ambitious rate-of-return goals, tolerating a significant level of risk and higher rates of portfolio turnover, conducting more frequent investment reviews, and maintaining a balanced portfolio weighted toward equities.

At the same time, the data also make clear that the vast majority of foundations, most of them small, continue to operate their investment function in a fashion closer to what *Institutional Investor* described in 1968: with only minimal trustee involvement, a variety of "satisficing" objectives, limited rate-of-return goals, low risk tolerance, a narrow time horizon, a less involved management style, and heavy reliance on fixed-income securities.

Judgments about the desirability of one approach compared to the other are no simple matter, of course. The jury is out on whether more frequent changes in portfolios provide better returns at lower risk than less frequent changes would achieve. Certainly, in the aftermath of the October 1987 "crash," it is hard to defend any particular investment approach as inherently superior.

Furthermore, like many of these issues, the appropriate approach depends upon the particular characteristics of any given foundation. A small foundation with a relatively short time horizon and a sense of limited longevity may be acting rationally in settling on a policy of maximizing income and not attempting to maximize capital over the long run. A large foundation that is run by a professional staff and that visualizes itself as operating in perpetuity must involve itself more in the investment management process; its officials cannot afford to ignore changes in the investment portfolio or to delegate major policy decisions to outsiders, as much as the expert help of outsiders may be useful.

Despite some evidence of considerable sophistication, the data presented here provide reason to suspect that in fact many foundations are giving insufficient attention to preserving and enhancing the available philanthropic resources under their control. The ultimate test of these suspicions, however, lies in the record of actual investment performance. Therefore, it is to this record that we now turn.

Chapter Four

FOUNDATION INVESTMENT PERFORMANCE

Ultimately, the real test of foundation investment operations is not how the process is structured but what the actual performance is. Indeed, the process is important precisely because it is expected to be an important determinant of performance.

Determining "what the actual performance is," however, is easier said than done. For one thing, as we have seen, there are a number of ways of defining investment performance—e.g., as the amount of investment income that is earned or as "total return" including both income and appreciation or depreciation in the value of the asset base. The latter is the more comprehensive approach, but it is also the more difficult since it requires data on the true market value of all assets and some method to take account of inflows and outflows of funds in the course of a year so that foundations are not unfairly credited with gains in their assets that were the result of gifts or faulted for reductions that were the result of generous grant making.

Beyond the problems of definition and data availability, performance figures may be affected more by general market conditions than by conscious investment decisions. Therefore, it is important to look not just at the raw performance figures but also at the investment climate in which they emerged and at the relationship between foundation performance and that of other similar institutions.

Finally, variations in performance can reflect variations in investment objectives. Some investors may choose lower performance because they are uncomfort-

able with the levels of risk that higher performance entails.

Because of these and other difficulties, solid data on the investment performance of foundations have long been lacking. A principal objective of our study was to help close this gap and develop a reasonably comprehensive overview of foundation investment performance covering a broad cross-section of foundations, not just the large ones, and spanning more than a single year.[1] Our goal, moreover, was to analyze not just foundation "income yield," but also the "total return" foundations were earning on their investments, including both income and appreciation. In addition, we sought to determine what impact, if any, the foundation investment practices and policies identified earlier actually have on

---

[1] In addition to avoiding having the results affected by the pecularities of any single year, the use of a multiyear time frame was designed to offset the effects of the fact that foundations do not all use the same fiscal year. For some foundations, the fiscal year begins on January 1 of each year and ends on December 31. For others, it begins on April 1 and ends on March 31. And for still others, other fiscal years are used. This difference can have significant consequences for the rate of return that is reported. For example, we conducted an experiment in which we assumed that we had twelve identical portfolios of stocks, bonds, and Treasury bills—each with a fiscal year starting on a different month—and then computed an annual rate of return for each in a given year. The resulting rates of return varied from a high of +15.62 to a low of −3.56. Computation of average rates of return over multiple years can even out these disparities and overcome a significant part of this problem.

foundation investment performance, and to assess the consequences for investment performance of the change in the payout requirement in 1981.

To explore these issues, the discussion in this chapter falls into four parts. The first section reviews key elements of our approach and puts the time period covered in this analysis into context. The second section then looks directly at the rate of return achieved by our sampled foundations, compares this to the return rate on a control portfolio, and assesses the impact on investment performance of the change in the payout requirement in 1981. In the third section, we attempt to explain the variations in foundation investment performance that we have found, focusing on many of the factors identified in Chapter Three, such as size, risk tolerance, management style, and asset composition. A final section summarizes the major findings and assesses their implications.

I.
THE APPROACH AND THE CONTEXT

The Data

The principal data sources used for our analysis of foundation investment performance were the 990-PF forms that foundations are required to file each year with the Internal Revenue Service. Although the 990-PF forms do not include a rate-of-return measure, they do contain enough information on the aggregate value of assets at the beginning and end of each year, and on gifts and expenditures, to make it possible to develop a reasonable approximation of each foundation's total rate of return after taking account of the inflow and outflow of funds. Accordingly, some twenty data elements were collected from the 990-PF forms available

on microfilm at the Foundation Center for the years 1979–1984 for each foundation we covered.[2]

The Sample

The sample utilized for this portion of our analysis was similar to that used for our survey work. In particular, form 990-PF data were sought on all 699 foundations with 1981 assets of $10 million or more, as well as on all 304 foundations with assets under $10 million that responded to our mail survey. This stratified approach was used to ensure sufficient coverage of foundations of all size classes and, particularly, to ensure adequate coverage of the larger foundations that account for the bulk of foundation assets. In addition, we wanted to make sure we could link the survey and investment data.

[2] The computation formula used was designed to approximate the "unit method" that investment analysts have developed to track investment performance in circumstances where there are extensive inflows and outflows from an investment fund. The "unit method," derived from mutual fund accounting methods, essentially treats a given portfolio of assets as a certain number of units available for investment for a given number of days. The investment return is then computed per unit. Because relatively few foundations use the unit method, and none are required to report in these terms on the 990-PF form, we developed an approach that is an approximation of the unit method. The formula we used was as follows:

$$R = \frac{(EFMV - BFMV) - Gifts + (Grants + Op.\,Exps. + Taxes)}{BFMV + Gifts/2}$$

Where:

EFMV = ending fair market value of assets
BFMV = beginning fair market value of assets
Grants = distributions made by the foundation
Gifts = contributions received by the foundation
Op. Exps. = operating expenses of the foundation

The formula assumes that grants, operating expenses, and taxes are paid out more or less as investment income is received, and that gifts are available for half of the year. See: J. Peter Williamson, "Investment Performance Measurement," *Financial Handbook.*

Table 4.1
Investment Data Sample

| Asset Size Class[a] | All Foundations | | Samples | | | |
| | Number | Percent | 990-PF | | 990-PF and Survey | |
			Number	Percent	Number	Percent
Large	166	0.7	88	19.2	45	22.0
Medium	533	2.5	252	54.9	77	37.5
Small	21,595	91.8	119	25.9	83	40.5
Total	22,294	100.0	459	100.0	205	100.0

[a] See Table 2.5.

Once account was taken of data availability, we were able to compile complete 990-PF data on 459 foundations for the full 1979–1983 time period. For 205 of these, we also had complete survey data as well.

As shown in Table 4.1, this sample contained enough representation in each of the size classes to permit valid generalizations about the universe of foundations in each class. In addition, by using weighting factors, it was possible for us to blow up the sample to the total universe of foundations.[3]

The Time Period and the Control Portfolio

Although we make repeated reference below to the investment environment of the 1979–1983 period, some preliminary observations are important to set the scene.[4]

To begin with, investors entered this period after a decade that had begun with the highest of expectations but that turned out to provide only turbulent and disappointing investment results. Consequently, investor expectations in early 1979 were far from optimistic.

Stock prices at that moment were no higher than they had been ten years earlier; they had spent most of the intervening time at even lower levels. Long-term interest rates were hovering at and were soon to break through historic highs, so that nearly every bond held was showing a loss.

Meanwhile, the cost of living had doubled. This meant that, with no gain or, more likely, an actual decline in nominal portfolio values, the real purchasing power of foundation capital had suffered catastrophic shrinkage. The income flow was little help. Dividend increases as well as Treasury bill income lagged behind the rate of inflation; bond interest obviously deteriorated in real terms. As a result, ambitious programmatic projections and commitments made in the late 1960s and early 1970s, to say nothing of legal minimum payout requirements, became increasingly difficult to finance.

Every five-year time period has its own investment peculiarities, and this one was no different. Most striking-

ingly, the years covered in this study bracketed a dramatic and essentially unexpected change in the economic environment. The inflation rate in 1979 exceeded 13 percent. In 1983, the inflation rate fell below 4 percent. The federal government deficit ballooned from $16 billion in 1979 to $176 billion in 1983 even while the trade-weighted foreign exchange value of the dollar, after three years of precipitous decline, zoomed upward by 50 percent.

Although no two years in this time period were alike in the capital markets, the results for the five years as a whole were considerably different from well-established, long-run rates of return for financial assets. Most significantly, even after taking account of the high rate of inflation that existed, especially in the early part of the period, the average annual rate of return from a composite portfolio of stocks, bonds, and bills was higher during this period than either the recent past (1950–1978) or the longer-term past (1926–1978) (see Table 4.2). This was due, first, to the unusually favorable rates of return on equities and, second, to the favorable return on Treasury bills. Bond returns were the only ones to fall below the rate of inflation during this period, and they fell below both their recent and longer term historic levels as well. Aside from the volatility that existed, therefore, 1979–1983 was a relatively favorable period for investment.

Table 4.2
Inflation-adjusted Annual Rates of Return on Assets, 1979–1983, versus Prior Periods

	Percent		
	1979–1983	1950–1978	1926–1978
Stocks	8.9	6.9	6.4
Bonds	−1.6	−0.5	1.5
Bills	2.7	0.1	0
Composite[a]	5.1	4.0	4.3

[a] 60 percent stocks, 30 percent bonds, 10 percent Treasury bills.

Nevertheless, it is important to emphasize that the investment results achieved during this time period deserve to be appraised in light of the rather disastrous performance and real losses that preceded them as well as in the light of the drastic transformation in the economic environment and topsy-turvy asset returns that accompanied them. For this reason, it is particularly important to look not at the raw numbers but at the relationship between foundation performance and the performance of other investment pools and market indexes, as we do throughout this analysis.

In particular, in an effort to factor out as much as possible the influence of external market forces, we developed a "control" portfolio for comparison purposes.

---

[3] Sample sizes of 50 or more are commonly considered necessary to support valid inferences in work of this sort. The overall sample, as well as all the separate subsamples except one (large foundations for which both 990-PF and survey data were available) met this minimum criterion, and the one that did not was quite close.

See Appendix A for a discussion of the weighting factors used and for the "standard errors" associated with the resulting estimates. Overall, we are 95 percent certain that our estimates of aggregate income yield are within 0.25 percent of the reported values and that our estimate of aggregate total return is within 0.12 percent of the reported value.

[4] Material on the investment climate was provided by Mr. Peter Bernstein, editor of *Institutional Investor*.

This portfolio seeks to replicate the results of a portfolio whose assets were distributed about the way the typical foundation's assets were distributed. This simulated portfolio was made up of 60 percent stocks, 30 percent fixed-income securities, and 10 percent cash. The rate of return for this fund was computed on the basis of the assumption that the stock portion of the portfolio performed at the same rate as the Standard and Poor's 500 stock index, that the bond portion performed at the same rate as the Salomon Brothers high-grade corporate bond index, and that the cash portion performed at the same rate as the Treasury Bill rate.[5] The result was a standard against which to measure the performance of our private foundation sample.

How, then, did foundations perform compared to this standard?

II.
OVERALL PERFORMANCE

Overall Return Rate for Foundations

As noted earlier, two basic measures of the rate of return on the investments of private foundations or other institutions are available. One of these—"income yield"—focuses exclusively on the annual income from investments, i.e., the interest and dividend earnings received and the realized appreciation from the sale of investment assets. The other—"total return"—includes not only this annual income on investments but also the unrealized appreciation or depreciation in the value of the investment assets still held in the investment portfolio.

Because the income yield measure ignores a major part of investment performance (i.e., growth or decline in the value of the asset base), the total return measure is preferred for the purpose of measuring the performance of an investment portfolio.[6] Yet, many of the prior studies of foundation investment return, including two very recent ones, have still utilized the income yield approach.[7]

Table 4.3 records the median compound annual rate of return for the foundations in our sample over the period 1979–1983, using both of these methods. The data recorded here have been weighted to represent the

[5] The asset composition of the control index was chosen to simulate the composition of the typical foundation portfolio. See: Williamson, "Inflation and the Foundation Payout Rate, 1981.

[6] The "income yield" measure also creates an artificial incentive for managers to put their resources into high-yield securities, which may not offer the best prospects for long-term growth. The "total return" approach, by contrast, credits an investment manager with good performance regardless of whether that performance shows up as growth in annual income from investments or as growth in the basic value of the investment portfolio itself.

entire universe of foundations, and the median value is reported so as to avoid the impact of large outliers on the results. In addition, the table reports both the actual value and the "real" value after adjusting for inflation.[8]

As this table makes clear, there is a considerable difference between the income-yield and total-return measures of investment performance. Based on the data we examined, the median foundation had an average income yield of 7.5 percent a year over this five-year period, and an average annual total rate of return of 12.4 percent a year.

Table 4.3
Foundation Investment Return Rates, 1979–1983

Return-Rate Measure	Median Compound Annual Rate of Return (Percent)	
	Actual	Inflation-adjusted
Income yield	7.5	−0.9
Total return	12.4	4.0

Source: Salamon/Voytek Foundation Investment Study.

With the inflation rate averaging about 8 percent a year during this period, the real, inflation-adjusted income yield for the median foundation was a negative 1 percent a year. The real, inflation-adjusted total return rate was 4.0 percent a year—i.e., assets grew at a rate that was 4 percent above the inflation rate. However, with a legal payout requirement of at least 5 percent, this means that the median foundation was being forced to deplete the real value of its assets during this period, as critics of the payout requirement have charged.

Variations in Rate of Return by Size of Foundation

The picture of foundation investment performance painted above tells only part of the story. Although the median provides an approximation of the rates of

[7] This is true, as noted above, of the analysis of 1982 990-PF forms conducted by the U.S. Treasury Department's Statistics of Income Division. It is also true of the General Accounting Office study completed in 1984. See: U.S. General Accounting Office, *Statistical Analysis of the Operations and Activities of Private Foundations* (Washington, D.C.: General Accounting Office, January 1984). One reason for this may be the 1969 payout requirement stipulating that all income must be paid out as grants, which focused attention on what this income was.

[8] The "median" represents the midpoint in a range of values. It is the value that lies right in the middle of all of those recorded, such that 50 percent of the observations are larger and 50 percent are smaller. Thus, the median is different from the "mean" or average, which represents the sum of all the observations divided by the number of observations. If some of the observations are way out of line with all the others, this will affect the mean but not the median. Thus, the median is a safer measure of the "central tendency."

return achieved by the "typical" foundation, it reveals little about the overall rate of return achieved on foundation *assets*. This is so, as we noted earlier, because foundation assets are not distributed evenly among foundations: A relative handful of the foundations control the vast majority of the assets. To the extent that rates of return differ systematically between large and small foundations, therefore, the foundation-weighted results reported above will differ from the asset-weighted results. This is so because the foundation-weighted results are most heavily driven by the more numerous small foundations, whereas the asset-weighted results are more heavily driven by the large foundations.

Table 4.4 makes clear that return rates do vary with the size of the foundation, although the differences are in opposite directions depending on which rate-of-return measure is used. In particular, the large foundations have a lower median *income yield* than the small- or medium-sized foundations, but they have a higher median *total rate of return* (14.9 percent versus 12.1 percent).

Table 4.4
Foundation Size and Return Rates, 1979–1983

Return Rate Measure	Median Annual Return Rate (Percent)		
	Large Foundations[a] (n = 88)	Medium Foundations[b] (n = 252)	Small Foundations[c] (n = 119)
Income yield	7.2	7.6	8.3
Total return	14.9	14.7	12.1

Source: Salamon/Voytek Foundation Investment Study

[a] $50 million and over.

[b] $10–49.9 million.

[c] Below $10 million.

Reflecting this, the asset-weighted *total return* rate was 14.1 percent a year, as shown in Table 4.5, compared to the foundation-weighted median of 12.4 percent. On the other hand, the asset-weighted *income yield,* at 6.3 percent, was below the foundation-weighted median income yield of 7.5 percent. These figures are consistent with the evidence on the different investment strategies of large and small foundations presented above. In the search for higher total rates of return rather than higher current income, the larger foundations tend to allocate a larger share of their assets to equities. And equities were the best-performing asset during the period 1979–1983.

Thanks to this performance by the larger foundations, the total return rate on foundation assets exceeded the inflation rate, producing a positive real rate of return on foundation assets of 5.7 percent, as shown

in Table 4.5. What this means is that the foundation community was able to maintain the real value of its assets and still sustain a 5 percent payout rate. However, many small foundations, with lower total-return rates, evidently lost ground.[9]

Table 4.5
Average Annual Asset-weighted Foundation Return Rates, 1979–1983

Return Rate Measure	Percent	
	Actual	Inflation Adjusted
Income yield	6.3	−2.1
Total return	14.1	5.7

Source: Salamon/Voytek Foundation Investment Study.

Comparison to Control Index

To interpret these results accurately, however, it is necessary to compare them to the performance of our control portfolio. This control portfolio, as pointed out above, consisted of 60 percent stock (S&P 500), 30 percent long-term, high-grade corporate bonds, and 10 percent Treasury bills.[10] Table 4.6 records the results of

[9] The return rates reported here—both the median and the asset-weighted rate—are considerably higher than the return rates that Professor Ralph Nelson recently computed for the 54 largest foundations over the period 1961–1982. According to Nelson, the estimated total return of these 54 largest foundations averaged 10.53 percent a year during the period 1961–1969, and 8.02 percent a year during the subsequent period, 1970–1982. By comparison, the asset-weighted total return for our sample was 14.1 percent for the period 1979–1983. Nelson, "Economic History of Large Foundations," p. 159.
  How can we explain this difference? Are foundations now doing a much better job of investment management, or are some other factors at work?
  A crucial part of the answer to this question, it turns out, lies in changes in overall market conditions. During the 1961–1969 and 1970–1982 periods that Nelson examined, the Standard and Poor's 500 stock index grew at an average annual rate of 8.70 percent and 7.94 percent, respectively. During the 1979–1983 period that is the focus of attention here, this same index grew at a rate of 17.3 percent—or twice as fast.

[10] These proportions were based on the asset composition records we were able to assemble on foundations, as recorded in Chapter 2, and on the advice of Dr. J. Peter Williamson, a consultant to this project, who pointed out that these proportions "reflect approximately a normal or average distribution of college and university endowment fund assets over the years, and therefore constitute probably the best estimate that can be made of the average distribution of foundation assets. These proportions are also approximately the average allocation of pension fund assets" (memo from J. Peter Williamson to Lester Salamon, October 20, 1986). Ralph Nelson suggests that a portfolio composed 80 percent of stock, 15 percent of bonds, and 5 percent of cash may be more appropriate for foundations (Nelson, "Economic History of Large Foundations," 1987, p. 159). However, Nelson bases this estimate on work he did in 1960, before the "excess business holdings" requirements of the 1969 Tax Reform Act were enacted.

Table 4.6
Comparison of Average Annual Total Return Rates for Foundations and Control Indexes, 1979–1983

Measure	Percent	
	Actual	Inflation-adjusted
1. Foundations		
A. Median foundation rate	12.4	4.0
B. Asset-weighted foundation rate	14.1	5.7
2. Control groups		
A. Standard and Poor's 500 stocks	17.3	8.9
B. Long-term bonds (Salomon Brothers)	6.8	−1.6
C. U.S. Treasury bills	11.1	2.7
D. Composite control group (weighted average of a, b, c)[a]	13.5	5.1
E. NACUBO	11.9	3.5
F. Common Fund[b]	16.1	7.7
3. Foundations versus composite control		
A. Median foundation as percent of control group		92.0
B. Asset-weighted foundation as percent of control group		104.0

Sources: Rows 1A and 1B from Salamon/Voytek Foundation Investment Study; rows 2A, 2B, and 2C from Ibbotson and Singuefeld, *Stocks, Bonds, Bills, and Inflation,* various editions; and as shown above.

[a] Assumes a portfolio composed 60 percent of stock, 30 percent of fixed income securities, and 10 percent of cash or cash equivalents.

[b] This is a composite of three separate Common Fund pools. See text for details.

this analysis. It compares both the median and asset-weighted total rates of return we computed for foundations for the 1979–1983 period to the total return rate for our composite control group over the same time period. Also noted for purposes of comparison are the rates of return on the separate components that comprise our control portfolio as well as two other comparison portfolios: first, the rates achieved by 250 college endowment funds as compiled by the National Association of College and University Business Officers (NACUBO), and, second, the return rate for a 60–30–

10 composite of three pools of stocks, bonds, and Treasury bills, respectively, managed by the Common Fund. Several observations flow from these data.

First, the composite control index outperformed the median foundation during this period. While the median foundation total-return rate during this period averaged 12.4 percent a year, the return rate achieved by the composite fund averaged 13.5 percent, a difference of 1 percent a year.

Second, once the foundation rates are weighted on the basis of assets, the relationship between foundation performance and the performance of our control fund is reversed. In particular, the asset-weighted total return rate for foundations, at 14.1 percent a year, was slightly higher than the 13.5-percent rate for the control group. In other words, while most foundations did not beat the control index, the larger ones, which account for most of the assets, did.[11]

Third, the performance of foundations was better than that of the NACUBO pool and worse than that of the Common Fund during this period, regardless of whether the median or asset-weighted foundation figure is used. Compared to the asset-weighted real foundation return rate of 5.7 percent, the inflation-adjusted rate achieved by the Common Fund was 7.7 percent.

The majority of private foundations thus performed worse than they would have if they had simply put their money into unmanaged broad stock and bond indexes with the asset mix of our control index. While the larger foundations have done better than this, producing a return rate on foundation assets that was higher than the broad market indexes, the difference certainly was not substantial. Among other things, these results seem to give credence to the "random walk" theory of investment management: Even with substantial management efforts and costs, foundations do not seem to be beating the broad market indexes by much.

How can we explain this pattern of results and what lessons does it hold? One explanation has already been hinted at: Most foundations may be performing

[11] Interestingly, this is similar to the result Nelson reports for his sample of the 54 largest foundations, at least during the 1970–1982 period. During the 1970–1982 period, Nelson found that his foundation sample beat a weighted average of stocks, bonds, and Treasury bills by 2 percent. However, Nelson assumed an 80–15–5 split among these components in constructing his index. If he had used a 60–30–10 split, the foundation advantage would have been 3 percent. During the earlier period he examined, 1961–1969, Nelson found that his large foundations beat the market averages by a larger margin. Nelson suggests that some of this difference may be attributable to the 1969 tax act, which induced foundations to diversify their holdings and build more conservative, income-oriented portfolios. Nelson, "Economic History of Large Foundations," pp. 159–160.

below our control portfolio because they have a smaller appetite for risk and therefore maintain a less aggressive asset mix than the 60–30–10 mix in the control portfolio or the other comparison groups. In other words, these foundations are implicitly choosing a lower return rate as the price of avoiding greater risk. This interpretation finds some support in the evidence presented earlier concerning the risk aversion, conservative investment goals, and relative inattention to the investment function among large numbers of foundations.

A second, more technical explanation for these results may be the fact that foundations must spend all or most of their income, thereby losing the opportunity to benefit from the magic of compound interest. By contrast, the control funds can reinvest their income and earn further returns on it. Over extended time periods, this can produce a notable difference.[12]

Impact of the Change in the Payout Requirement

The results reported above treat the 1979–1983 period as a single unit, but let us now divide this period into two subperiods, separated by the change in the payout requirement in 1981. This step provides an opportunity to assess whether this change, which eliminated the requirement that foundations must pay out all of their investment earnings if they exceeded 5 percent of assets, had any impact on foundation investment performance. We can respond then to the argument made by some observers that the old payout requirement tended to inhibit foundations from investing in high-yield securities during inflationary periods because it required them to pay out all of their income. To what extent do the data seem to confirm this argument?

---

[12] A portfolio with an income return of 5 percent a year that is reinvested to earn 5 percent a year would accrue a total of 27.5 percent over a five-year period, or 5.5 percent a year. By comparison, a portfolio that spends the full 5 percent would earn 25 percent over the five-year period.

Although it is difficult to reach firm conclusions on the basis of the limited two-year, post-1981 period for which we have data, several interesting conclusions emerge from our recalculation of foundation return rates for the two subperiods, 1979–1981 and 1982–1983.

In the first place, both the median and the asset-weighted foundation return rates are higher for the post-1981 period than they are for the pre-1981 period, even before taking inflation into account. The median foundation total return went from 10.0 percent a year in the earlier period to 14.9 percent a year in the latter period while the asset-weighted foundation return went from 12.8 percent to 16.0 percent. This suggests that foundations performed much better after the change in the payout requirement than they did before (see Table 4.7).

These disparities are even more striking once the differences in inflation rate between these two periods are taken into account. During the 1979–1981 period, inflation averaged 11.5 percent per year. During the 1982–1983 period, by contrast, inflation averaged 3.8 percent per year. Expressed in real, inflation-adjusted dollars, therefore, the *median* foundation rate of return went from − 1.5 percent a year in the earlier period to + 11.1 percent a year in the latter period. Adjusted for asset size, the real foundation rate of return went from + 1.3 percent a year to + 12.2 percent.

To what extent was this shift a product of the change in the payout requirement and to what extent was it merely a reflection of what was happening in the capital markets? Our control portfolio can help us to answer this question.

As it turns out, the annual rate of return on the control portfolio rose even more dramatically than that for the foundation sample — from 8.8 percent in the earlier period to 20.9 percent in the latter period. What this suggests is that the increase in foundation return rates in the period since 1981 was probably due more to changes in market conditions than to changes in the payout requirement. This conclusion finds further

Table 4.7
Comparison of Total Return Rates for Foundations and Composite Control Group Before and After 1981 Payout Change

Time Period	Annual Compound Total Rate of Return (Percent)			Foundations as Percent of Control Group	
	Foundations		Control Index		
	Median	Asset-weighted		Median Foundation Rate	Asset-weighted Foundation Rate
1979–1981	10.0	12.8	8.8	114	145
1982–1983	14.9	16.0	20.9	71	77
Total, 1979–1983	12.4	14.1	13.5	92	104

Source: Salamon/Voytek Foundation Investment Study.

confirmation in our survey results. Only about a third of the foundations represented in our survey indicated that the payout requirement was either "extremely important" or "fairly important" to their investment performance during this period.

But why did foundations not benefit as much from the changed market conditions as the market averages? Why did their performance not improve as much as the composite index?

The data presented earlier suggest a possible answer to this question. As we have seen, foundations tend to be risk-averse investors: They tend to shy away from volatile investments, such as stocks and long-term bonds. During periods of market decline, this strategy can produce above-average returns. But during periods of robust growth, this strategy is likely to produce below-average returns.

Since 1979–1981 was a weaker period than 1982–1983, it is understandable why foundations would beat the composite index in the earlier period but then fall behind the index more recently. Annual rates of return on stocks, for example, swung from 14.2 percent during 1979–1981 to 22.0 percent during 1982–1983; bonds made an even bigger swing from −2.6 percent to +22.7 percent, outperforming stocks; paper, on the other hand, fell from an extraordinary 12.1 percent to a still-high 9.7 percent.

Most foundations likely held less than 60 percent of their portfolios in stocks and also held more short-term paper and fewer long-term bonds than the 10–30 allocation in our control portfolio. Consequently, they missed out both on the huge upswing in the long-term bond market and the considerable improvement in returns on stocks.

The finding that the change in the payout requirement had far less influence on performance than did shifts in market conditions may have been affected as well, however, by the fact that too little time had elapsed between the change in the payout requirement and the period for which data are reported in Table 4.7, i.e., 1982–1983. To check this, we extended our analysis to 1984 for the more limited number of foundations for which 1984 data were available. Focusing only on foundations with a December fiscal year to avoid the fiscal year problem, we discovered that the median total rate of return for these foundations in 1984 was 8.4 percent, down considerably from the earlier period.[13]

[13] To make sure the apparent drop in investment performance in 1984 was not because we were dealing only with foundations that were weak performers, we also computed the 1979–1983 return rate for them. As it turns out, the unweighted return rate for these foundations during 1979–1983 is very close to that of the entire sample. For all of these 1984 calculations, only foundations with a December fiscal year were used to avoid the problems of varying fiscal years.

Nineteen eighty-four was a year in which previously high stock returns fell to only 6.3 percent, lagging far behind the 16.4 percent return on bonds and even the 9.9 percent return on Treasury bills. In a reflection of this, the rate of return on our control portfolio was 9.7 percent, down from 20.9 percent during 1982–1983.

In other words, the control portfolio outperformed the foundation sample in 1984, even though this was a down year for the equity markets, and past experience suggested that foundations do better relative to a 60–30–10 control portfolio during such periods. Although the 1984 sample is too limited to read too much into these results, there is at least a hint in them of a greater willingness to take risks on the part of foundations. Because this was a down year for the equity markets, however, this may have depressed returns somewhat, leaving the foundations with a lower rate of return than the control portfolio.[14]

III.
EXPLAINING FOUNDATION INVESTMENT PERFORMANCE

The findings on overall foundation performance presented above relate to the foundation universe as a whole. In fact, however, rates of return varied widely among foundations. Fifty-two percent of the 3,775 observations of foundation return rates in our data set for the period 1979–1984 were observations in which foundations did better than the control portfolio, and 48 percent were observations in which foundations did worse (see Table 4.8). As with the overall return rate,

Table 4.8
Comparison of Foundation Investment Performance to the Composite Index, 1979–1984

Year		Percent of Foundations with Annual Total Rate of Return	
		Above Composite Index	Below Composite Index
1979–1981	(n = 2,153)	58.0	42.0
1982–1984	(n = 1,622)	44.1	55.9
1979–1984	(n = 3,775)	52.1	47.9

Source: Salamon/Voytek Foundation Investment Study.

[14] One other plausible explanation of the 1984 results is that the 1984 sample contains a higher proportion of larger foundations, which, as we saw earlier, tend to invest more heavily in equities. The relatively low 1984 performance of foundations may therefore be the result not of the payout requirement change but of the inclusion of a larger number of these less-risk-averse foundations.

moreover, a larger proportion of foundations beat the control portfolio in the earlier portion of this time period (1979–1981) than in the later portion (1982–1984), as Table 4.8 also shows.

How can we explain these variations? What clues do they hold about the causes of successful or unsuccessful investment performance by foundations? What impact, if any, have the investment policies and practices examined in Chapter Three had on investment performance?

One part of the answer to these questions has already been explored: Foundation *size* seems to have some impact on investment performance. As Table 4.4 showed, the median total return rate for large foundations (those with assets in excess of $50 million) was 14.9 percent over the 1979–1983 period. In contrast, the median total return rate for small foundations (those with assets of $10 million or less) was 12.1 percent.

These facts alone do not provide enough evidence for us to conclude that size is the primary determinant of differences in foundation performance, however. Indeed, 43 percent of the large foundations had rates of return that lagged behind the control portfolio while 46 percent of the small foundations exceeded it. Clearly, something more than size seems to be at work here.

As the discussion in Chapter Three made clear, foundation size is associated with a number of other facets of foundation investment behavior, such as investment objectives, use of outside management, risk tolerance, and asset composition. To what extent are these other factors, rather than size alone, responsible for the apparent disparity in performance between large and small foundations? And what implications does this have for the management of the foundation investment function and for our understanding of investment management more generally?

To answer these questions, we drew on both our survey and 990-PF data to examine the relationship between foundation investment performance and the investment strategy and process concerns explored in Chapter Three.[15] This section first examines the apparent relationship between investment performance and some of these major factors taken individually; it then attempts to sort out the relative influence of each using a method known as "multiple regression analysis."

---

[15] Because not all the foundations for which we gathered valid investment return data completed our survey, the number of observations for this portion of the analysis is smaller than for the earlier discussion of investment rate of return. In particular, compared to the 459 foundations for which we have valid form 990-PF data for the 1979–1983 period, the discussion in this section draws on a sample of 205 foundations for which both 990-PF and survey data are available. Inevitably, the inferences that can be drawn from this more limited sample must be more tentative.

Rate-of-Return Objective and Investment Performance

One of the first factors worth exploring is the kind of investment objectives foundations are pursuing. Larger foundations may have performed better than smaller ones during this particular period because they tend to focus more heavily on total return as their principal investment objective, leading to a heavier concentration in the best performing asset in this period—common stocks.

According to this line of argument, foundations that focus on total return are more likely to achieve a higher total return rate over the long run than those that seek to maximize only income yield, which is one component of total return. At the same time, however, total-return policies tend to produce a more volatile pattern of returns and can therefore lead to poor short-run results.

The relationship between investment objective and return rate is also relevant to the whole issue of the impact of the 1981 payout requirement on foundation investment performance. One of the arguments for this change was that it would free foundations to focus on a total-return objective instead of focusing on policies to minimize current-income yields.

To what extent do the data reveal a relationship between investment objective and realized rates of return?

The answer, shown in Table 4.9, is that there does seem to be a relationship between the investment objective of a foundation and the rate of return it achieved during the years 1979–1983. The median total rate of return for foundations that defined their investment objectives in terms of maintaining program spending levels, maximizing investment income but not appreciation, or maintaining the real value of their assets was 12.5 percent or less over the 1979–1983 period. In comparison, the median return rate for foundations that defined their investment objectives in terms of cap-

Table 4.9
Relationship Between Foundation Investment Objectives and Investment Performance

Objective	Median Compound Annual Total Rate of Return (Percent)
Maintain program spending level ($n = 14$)	9.4
Income maximization ($n = 24$)	11.9
Maintain real asset value ($n = 26$)	12.5
Capital appreciation ($n = 10$)	13.5
Total return ($n = 130$)	14.0

Source: Salamon/Voytek Foundation Investment Study.

ital appreciation or maximization of total return was noticeably higher, at 13.5 percent or 14 percent.

Risk Tolerance and Investment Return Goals

If foundation investment performance appears to have been affected by the type of return goal foundations were seeking, it also seems to have been affected by the level of risk they were willing to tolerate. As Table 4.10 shows, the higher the foundation's risk tolerance, the greater the return rate it achieved, other things being equal.[16]

Table 4.10
Relationship Between Risk Tolerance and Investment Performance

Risk Tolerance[a]	Median Compound Annual Total Rate of Return (Percent)
Very low	12.9
Moderately low	13.0
Moderately high	13.7
Very high	14.7

Source: Salamon/Voytek Foundation Investment Study.
[a] See Table 3.3 for definitions.

Similarly, foundations with higher rate-of-return goals generally performed better in this period than those with lower goals. We can see these differences in Table 4.11, which shows that foundations with three-year rate-of-return goals of 13 percent or more had an actual median rate of return of 15.3 percent per year, compared to 12.8 percent for the foundations with lower goals.

Table 4.11
Relationship Between Foundation Rate-of-Return Goal and Investment Performance

Three-year Rate-of-Return Goal	Median Compound Annual Total Rate of Return (Percent)
6–9 percent	12.8
10–12 percent	13.8
13 percent and over	15.3

Source: Salamon/Voytek Foundation Investment Study.

It is important to keep in mind that these results may be time-specific. During 1979–1981, for example, portfolios without an ample cushion of short-term paper or with an excessive weighting in long-term bonds would have produced bad results that only the events of 1982–1983 would have been able to reverse. In other words, our data do not prove that aggressive investment policies will produce consistently better performance. Riskier portfolios by definition will have greater variation in the short run. But these data do suggest that such portfolios will perform better over the long run, and this is a significant finding. What it suggests is that differences in attitudes toward risk or return goals do seem to produce differences in realized rates of return.

Portfolio Composition and Investment Performance

That risk tolerance and rate-of-return objective affect foundation performance is probably due to the effect they have on the asset mix. Although the fit is far from perfect, since both stocks and bonds vary in their degree of riskiness, foundations with higher risk tolerance and longer time horizons will tend to select portfolios with larger proportions of common stocks, while those with less risk tolerance are more likely to invest more heavily in fixed-income securities.

Although our data on asset mix have serious limitations, they nevertheless permit us to draw some suggestive conclusions about the influence of asset mix on performance.

Table 4.12 records the performance of foundations with different shares of their assets in common stock.[17]

Table 4.12
Relationship Between Foundation Asset Composition and Investment Performance, 1979–1983

Percentage of Assets in Corporate Stocks	Median Compound Annual Total Rate of Return (Percent)
60 or more	16.0
40–59	13.3
Under 40	12.9

Source: Salamon/Voytek Foundation Investment Study.

[16] As noted more fully in Chapter Three, very low risk tolerance was defined for the purposes of this question as "less than the popular bond market averages," moderately low risk tolerance as "about equal to popular bond market averages," moderately high risk tolerance as "about equal to the popular stock market averages," and very high risk tolerance as "greater than the popular stock market averages."

[17] As noted in the methodology section above, the 990-PF form that is the basis for these data stopped collecting separate information on equities (stock) versus fixed-income securities in 1981 and grouped the two together in one category, called "securities." The data in Table 4.12 therefore use the asset mix as of 1980. The argument for using these data is that foundations that were relatively heavily invested in stocks in 1980 were also likely to be relatively more heavily invested in them throughout the entire period. But this limitation of the data should be taken into account when evaluating these results.

What it shows is that foundations with 60 percent or more of their assets in stocks had a median rate of return during 1979–1983 that was significantly higher than that of foundations with less than 40 percent of their asset in stocks. This is consistent with our previous findings about the relationship between risk tolerance and performance, although these results might have been entirely different had equities not been the best-performing asset in this time period.

Outside Management and Investment Performance

A direct relationship also seems to exist between the number of outside managers a foundation uses and its rate of return. Table 4.13 shows that foundations with no outside managers or only one outside manager recorded a median rate of return close to 13 percent between 1979 and 1983. By contrast, the median return for those that made use of multiple outside managers was close to 15 percent or better. This difference may be less a reflection of the work of the managers, of course, than a more general reflection of the overall investment philosophy of the foundation. For example, foundations employing multiple outside managers were also more likely to have a higher risk tolerance and larger portions of their portfolios invested in equities, which performed well during this period.

Table 4.13
Relationship Between the Number of Outside Managers and Foundation Investment Performance

Number of Outside Managers	Compound Median Annual Total Rate of Return (Percent)
None	12.8
1	13.0
2–3	14.7
4 or more	15.8

Source: Salamon/Voytek Foundation Investment Study.

On the other hand, a much weaker relationship exists between foundation return rates and the *percentage* of foundation assets managed externally. The median annual return rate for the foundations with none of their assets managed externally was 12.8 percent between 1979 and 1983, while the rate of return for the foundations with two-thirds or more of their assets managed externally was 13.5 percent, only slightly better. One probable reason for this small difference is that a substantial number of smaller foundations have essentially turned the investment function over to a single outside manager, but with little incentive or directive to manage the funds very aggressively. Were these more conservative, outside-

managed funds excluded, a stronger relationship might emerge between the extent of reliance on outside management and the rate of return.

Management Style and Investment Performance

Consistent with the thrust of the findings presented above, foundations that took a less active approach to managing their assets had a moderately lower rate of return during 1979–1983 than the foundations that utilized a more active style. Table 4.14 shows the median rates of return over this time period for foundations that reported each of the three investment management styles identified in our survey—passive (or relatively inactive), moderately active, and active. The relationship here is far from perfect because the most actively managed foundations did slightly worse than the moderately active ones. Nevertheless, there does seem to be a general tendency for the rate of return among relatively actively managed portfolios to be higher than the returns among the less actively managed portfolios.

Table 4.14
Relationship Between Management Style and Investment Performance, 1979–1983

Management Style[a]	Compound Median Annual Total Rate of Return (Percent)
Passive	12.4
Moderately active	13.9
Active	13.1

Source: Salamon/Voytek Foundation Investment Study.

[a] See Chapter Three (page 00) for definitions.

This pattern finds additional support in the data relating return rates to the overall investment management "scores" computed for each foundation in Chapter Three. Table 4.15 shows that during 1979–1983 foundations with the highest such scores, indicating a high rate-of-return goal, use of a total-return objective, high risk tolerance, broad time horizon, and

Table 4.15
Relationship Between Foundation Investment Management Score and Investment Performance, 1979–1983

Investment Score[a]	Compound Median Annual Total Rate of Return (Percent)
Low	12.7
Medium	13.6
High	15.2

Source: Salamon/Voytek Foundation Investment Study.

[a] See Chapter Three (page 00) for definitions.

active management style generally had a higher median rate of return than did foundations with low scores.

Summary: The Regression Results

The data presented here thus suggest that the investment policy and practice factors identified in Chapter Three did have a significant impact on foundation investment performance. Active and aggressive investment management does seem to have resulted in significantly higher rates of return than less aggressive approaches, *at least during the period covered here.*[18]

Before we rest content with this conclusion, however, let us take the analysis one step further. The close interrelationship among the various factors involved in investment management may have led us to attribute to one factor what in fact was the result of another. For example, if both foundation size and asset composition are closely related to investment performance but are also closely related to each other, how do we know whether it is size or asset composition that is really the cause of higher performance? Conceivably, moreover, some third factor may be at work that affects the other two. It is necessary, therefore, to sort out the relative influence of the various factors thought to be affecting investment performance and to determine what share of the overall variation in return rates each one accounts for.

The best way to accomplish this is to use a statistical technique known as "multiple regression analysis," which essentially computes the independent effect of each of a variety of factors thought to be causing variation in a given "dependent variable" while holding the other factors constant. Regression analysis enables us to determine three things: first, the direction of the relationship between each factor and investment performance, with all the other factors held constant; second, the size of the change in investment performance that is associated with a given change in each factor; and third, the probability that the given result could have happened purely by chance rather than as a result of a true causal link. Generally speaking, only results that are not likely to have happened by chance more than one in twenty times are considered "significant" rather than random.[19]

Table 4.16 reports the results of such a regression analysis relating the factors identified above to foundation investment performance. The left-hand column in the table lists each of the factors that we thought likely to influence investment return. The right-hand column provides three separate pieces of information—whether the relationship between the factor and rate of return was negative or positive, how large a change in the rate of return was associated with a one-unit change in the factor, and whether the relationship was statistically significant (indicated by an asterisk).

Table 4.16
Factors Affecting Variations in Foundation Investment Performance, 1979–1983: Regression Results ($n = 211$)

Factor	Regression Coefficient
Size (asset value, 1982)	.12
Management style	−.21*
Formal review process	.30*
Time horizon	−.11
Risk tolerance	.02
Rate-of-return goal	−.06
Investment objective	−.02
Age	.15
Share of assets managed externally	−.07
Number of managers	−.07
Percent of assets in stock	.09
Percent of assets in fixed income	−.30*
Percent of assets in cash, savings	.24*
Total share of variation explained: 18 percent	

Source: Salamon/Voytek Foundation Investment Study.
* Significant at the .05 level.

Two surprises are immediately apparent in these regression results. The first is that foundation size, although positively related to performance during this period, as predicted, was nonetheless not a statistically significant determinant of differences in rate of return.[20] Hence, the size-related data we examined earlier may be only a proxy for other determinants.

A second surprise was that time horizon, rate-of-return goal, investment objective, the share of assets managed externally, and the number of outside managers had negative rather than positive relationships to return. The relationships were not statistically significant, however, so that the basic conclusion to draw from the data is that these factors do not seem to have any demonstrable independent effect on performance.

---

[18] Beyond the implications it has for foundation investment management, this conclusion also seems to run counter to the "random walk" theory of market behavior, which essentially holds that over time it is not possible to beat the market averages through effective management of investments. Although the "random walk" theories apply chiefly to the choice of individual securities within a particular asset class rather than the choice of asset mix, they nevertheless suggest an outcome different from that reported here.

[19] It is well to remember that regression analysis measures statistical relationships among variables. It does *not* distinguish cause from effect, although we may use its results to support a hypothesis that the relationship runs from one variable to another.

[20] More precisely, the observed result might have happened by chance more than one in twenty times.

Four factors do show a statistically significant association with foundation rates of return, however. Two of these are essentially measures of asset mix—percent of assets in fixed-income securities and percent of assets in cash and cash equivalents. The other two are measures of the degree of activism in the management of investments—the presence of a formal review process and management style (active, somewhat active, and inactive).

As we might expect, a significant negative relationship existed between the share of assets in fixed-income securities and the rate of return. In other words, the higher the share of assets in fixed-income securities, the lower the rate of return. This is consistent with the fact that common stocks earned more than twice as much as bonds and half again as much as Treasury bills during the years from 1979 to 1983.

On the other hand, the percent of assets held in cash or cash equivalents showed a significant positive relationship to rate of return. This result also reflects the peculiarities of the period, in which fixed-income portfolios with short maturities outperformed long-term bonds by a wide margin, especially during 1979–1981.

The existence of a formal review was also related to return in a significant and positive fashion. This suggests that foundations that actively and regularly review their asset mix in light of changing market conditions will tend to do better than foundations supervised in a more casual manner.

The significant negative relationship between rate of return and investment style—more or less activism—is more difficult to interpret. We surmise that the asset mix decisions are dominant, with the rate of turnover in holdings and the amount of management effort expended secondary in impact, or actually negative, once the basic asset mix is set.

These findings must, of course, be viewed with a degree of caution in view of the limited number of observations on which the regressions could be computed, the peculiarities of the time period, and the high degree of correlation among some of the variables (what statisticians refer to as "multicolinearity"). Furthermore, the regression equation we have identified accounts for only 18 percent of the variance in foundation performance—a significant share for a process as complex as the one we are examining, but still not sufficient to account for all of the variation.

Nevertheless, the results do provide some very useful insights into what distinguished one pattern of foundation performance from another in a difficult environment for investment management. Once controls are introduced for many of the other factors potentially influencing performance, the independent effect of the key process and management factors examined above—such as time horizon, rate-of-return goal, and investment objective—largely disappears. The same is true, moreover, of the size factor. What emerges instead as the principal determinant of foundation rates of return is the degree of risk a foundation is prepared to accept, as reflected in its asset allocation decisions and the presence of procedures for reviewing asset composition on a regular basis. Interestingly, this is precisely the conclusion that capital market theory predicts—i.e., that risk is the critical variable in the relative performance of investment portfolios. We can put it another way by saying that our results support the theory that differences in risk explain far more than differences in skill in explaining why one portfolio's rate of return is higher or lower than another's.[21]

IV.
CONCLUSIONS

Five major conclusions flow from this analysis of foundation investment performance.

First, aggregate foundation assets earned an overall rate of return during the years 1979–1983 that was generally respectable in that—over the period as a whole—it did not deviate widely from the broad market averages. In fact, the total rate of return on assets was slightly above the return on a simulated control portfolio composed of 60 percent stock, 30 percent high-grade, long-term corporate bonds, and 10 percent Treasury bills. Indeed, the total return was nearly 6 percent above the rate of inflation during this time period. This was just about enough to finance a 5-percent payout rate and operating expenses without invading the real capital of the foundation asset pool. By the same token, however, real capital essentially failed to grow over these five years.

Second, the real return of nearly 6 percent was the result of very high returns during the last two years of the period as inflation tapered off and the capital markets responded favorably. The real return was clearly negative during 1979–1981, which would have meant a severe impairment of real purchasing power had not the good fortunes of 1982–1983 arrived so soon after the earlier agonies.

Third, this record of performance is an asset-weighted result, skewed heavily upward by the higher returns earned by the larger foundations that tended to be willing to take larger risks. Foundations with less aggressive and less flexible investment objectives—typically the more numerous smaller foundations—probably held too much of their assets in cash equivalents to

---

[21] The literature is replete with the development of this basic idea. See William F. Sharpe, *Investments,* 3rd Ed. (Englewood Cliffs, N.J.: Prentice-Hall, Inc.). See Chapter 7 for a detailed and authoritative summary of these concepts.

share fully in the revival after 1981. As a result, most foundations performed below the simulated control portfolio.[22]

Fourth—a conclusion implicit in the three already set forth—the foundation investment world is far from homogeneous, which, together with the peculiarities of this time period, means that generalizations about it must be interpreted with appropriate caution.

Finally, the data presented here provide some support for a central tenet of the "random-walk" theory of investment management—i.e., that the best, and possibly only, way to boost investment return is to accept greater risk. In doing so, however, these results may help to explain why the median foundation return rate was not higher than it is. Higher risk, after all, is a two-edged sword. While it permits greater returns when market conditions are good, as they generally were during the period covered here, it also produces greater

losses when markets turn sour, as they have in the past and likely will at some point in the future. Unlike pension funds, which have steady inflows of cash, foundations have no other sources of income and must therefore finance their current operations almost entirely out of current earnings on assets. As a result, they have good reason to moderate the risk they assume and behave somewhat more conservatively than other institutions. This is particularly true given the terms of the payout requirement, which, even after the 1981 change, still ties the payout level to a given year's asset value rather than to a floating average of several years. It is reasonable to assume that this places a limit on the time horizon foundations can reasonably use in setting their investment objectives and presumably limits the risk they can afford to take. Thus, while there is probably solid room for improvement in foundation investment performance, there is also a solid basis for understanding why most foundations underperform the averages.

This final conclusion, however, shows how important the linkages are between investment policy and payout policy. Therefore, we now turn to an examination of that very issue.

---

[22] As we will see more fully in the next chapter, these risk-averse foundations also unfortunately tended to have high current income yields and therefore were forced to have the highest payout ratios prior to 1981, just when the inflationary assault on the real value of foundation capital was at its most intense.

Chapter Five

FOUNDATION PAYOUT POLICIES AND PERFORMANCE

Foundation investment management is not an end in itself. Rather, it is a means to enable foundations to carry out their charitable missions, to distribute the resources in their control for charitable purposes. But critical choices have to be made about how to balance current and future beneficiaries of foundation largesse. Other things being equal, the higher the current level of payout is, the smaller is the pool of assets left to support future distributions. On the other hand, the more investment earnings that are retained and plowed back into the asset base, the more questionable is the foundation's claim to tax-exempt status as a charitable institution. The task of foundation financial management therefore goes beyond merely earning a high rate of return; it involves striking a meaningful balance between these competing concerns.

Prior to 1969, as we have seen, foundations were left essentially on their own in making these decisions, albeit with some general congressional guidance after 1950 about the need to avoid excessive accumulations of resources. In response to concerns that at least some foundations were abusing this freedom and neglecting to distribute sufficient funds for charitable purposes, however, Congress changed this situation fundamentally in 1969, enacting a "payout requirement" that established a mandatory floor for foundation charitable distributions. Henceforth, foundations had to pay out each year in "qualifying distributions" (grants plus related taxes and administrative expenses) the higher of

(a) their total investment income;[1] or (b) their "minimum investment return," essentially a percentage of their investment assets, a percentage that has been set since 1976 at 5 percent.[2]

In 1981, Congress agreed to liberalize this payout requirement by eliminating the first of these two provisions. Foundations must still distribute at least 5 percent of assets, but they are no longer required to pay

---

[1] The requirement was actually to pay out all of "adjusted net income," which includes gross income from all sources, including tax-exempt income, income from short-term capital gains, interest and dividend income on investment, any amounts received from repayments of prior qualifying distributions, and any released set-asides—minus deductions that ordinarily would be allowed to a regular taxable corporation.

[2] For purposes of computing the "minimum investment return," investment assets are all assets other than those used directly for the foundation's charitable purposes (e.g., a building in which the foundation carries out its charitable activities). Also excluded is acquisition indebtedness related to such directly used assets. The base for the computation of the minimum investment return is a floating average of the monthly total market value of the foundation's investment assets. The minimum investment return was originally pegged to a formula based on various economic indicators. In 1969, this led to a rate of 6 percent. In 1976, the idea of a fluctuating rate was dropped, and the law was changed to fix the minimum investment return percentage at 5 percent of investment assets. For more detail on the definition of "minimum investment return," see: Laurens Williams, "Definitions, Excise Tax, and Distribution Requirements," in *Foundations and the Tax Reform Act of 1969* (New York: The Foundation Center, 1970), pp. 8–9.

out income they earn in excess of that amount. This change was made in response to complaints that the old rule prevented foundations from reinvesting a portion of their earnings and was contributing to an involuntary decline in the value of foundation assets in an inflationary environment. Beginning in 1982, foundations were allowed to retain earnings in excess of 5 percent of their assets.

Previous chapters have examined the impact of this 1981 change on the *investment* behavior of foundations. This chapter turns to the impact on *payout* behavior.

This latter impact is particularly important in view of the fact that one of the great concerns about the 1981 change was that it might lead to a significant decline in payout rates. To what extent has this concern proved warranted? To what extent is it likely to prove warranted in the foreseeable future? Which foundations are most likely to be affected by the 1981 change?

To answer these questions, the discussion in this chapter falls into four sections. Section I sets the context by examining the foundation payout record over the period 1979–1983 and exploring the relationship between payout performance and investment performance. As we shall see, payout performance and investment performance turn out to be inversely related to each other—i.e., the higher the return rate, the lower the payout rate. Section II seeks to unravel this paradox, looking particularly at variations in how foundations approach their payout decisions and at the relationship between payout rates and investment strategy. Section III then analyzes the impact that the 1981 change in the payout requirement has had to date on payout rates and speculates, on the basis of the earlier findings, what the impact is likely to be in the near future. A concluding section then spells out some of the implications of these findings.

I.
THE OVERALL RECORD OF FOUNDATION PAYOUT PERFORMANCE

The 1969 payout requirement essentially established a floor on the ratio of foundation payout levels to foundation assets. It did not, however, prevent foundations from functioning above this floor. Therefore, it is important to examine at the outset how foundations performed in relation to this floor and what variations in this performance are apparent.

Overall Payout Performance, 1979–1983

Perhaps the major conclusion to emerge from the data on foundation payout performance during 1979–

1983 is that foundations achieved a fairly respectable payout rate over this period. Compared to a minimum-payout rate of 5 percent of assets set by law, the median foundation represented in our sample recorded an average annual payout rate of 7.8 percent during the five years between 1979 and 1983.[3]

Interestingly, however, while the payout *rate* was well above the minimum, the actual amount of "qualifying distributions" did not grow in real terms over this period. As noted in Table 5.1, qualifying distributions increased by 38 percent over the five-year period covered here, but this was just barely enough to keep up with inflation. In real, inflation-adjusted terms, the qualifying distributions of the foundations that we studied increased by only 1 percent over this five-year period. In comparison, the value of the investment assets of these foundations increased by 13 percent in real terms.

Table 5.1
Growth in Qualifying Distributions and Investment Assets of Foundations, 1979–1983

	($ Billions)		Percent Change, 1983 versus 1979	
	1979	1983	Actual	Inflation-adjusted
Qualifying distributions	2.6	3.6	+38	+ 1
Investment assets	39.5	59.4	+50	+13

Source: Salamon/Voytek Foundation Investment Study.

Foundation Size and Payout Rate

One important reason for this outcome is that not all foundations had payout rates at or near the median. Rather, payout rates tended to vary inversely with the size of the foundation. This is evident in Table 5.2, which records the median payout rates of foundations of different sizes over the 1979–1983 period. This table shows that the largest foundations, which control the bulk of foundation assets, had a median payout rate in the neighborhood of 6 percent a year over this period. The more numerous smaller foundations, on the other hand, had a median payout rate in the neighborhood of 8 percent. As a result, while the median foundation payout rate over the 1979–1983 period was 7.8 percent, the asset-weighted rate was a more modest 6.8 percent.

---

[3] The payout rate equals the amount of "qualifying distributions" divided by the average investment assets.

Table 5.2
Foundation Payout Rates, by Size of Foundation, 1979–1983

Group	Average Annual Payout Rate 1979–1983 (Percent)
Large foundations[a]	6.0
Medium foundations[b]	6.6
Small foundations[c]	8.0
All (median)	7.8
Asset-weighted	6.8

Source: Salamon/Voytek Foundation Investment Study.
[a] Assets of $50 million or more.
[b] Assets of $10–49.9 million.
[c] Assets of under $10 million.

Investment Performance and Payout Performance

The lower payout rates of the larger foundations raise important questions about the relationship between payout rates and rates of return on investment, because the larger foundations as a group had higher rates of return than the universe as a whole during 1979–1983. In fact, as the data in Table 5.3 show, there is a negative correlation between payout level and investment return for each of the years we examined except for 1979. Furthermore, this inverse relationship was particularly significant during 1981–1983. In other words, the higher the rate of return, the lower the payout rate.

Table 5.3
Correlation Between Payout Rate and Investment Return Rate, 1979–1983

Year	Correlation Coefficient
1979 ($n = 669$)	+0.04598
1980 ($n = 668$)	−0.01617
1981 ($n = 720$)	−0.11941*
1982 ($n = 733$)	−0.10879*
1983 ($n = 591$)	−0.11944*

Source: Salamon/Voytek Foundation Investment Study.
*Significant at the 0.05 level.

One explanation for this inverse relationship may be that foundations were taken by surprise by the bull markets of 1982–1983, with grants inadvertently lagging behind. This would most likely be the case with those foundations that had the greatest appreciation in their portfolios. The fact that payout decisions are based on investment performance in the prior year may help explain this.

We must, however, probe more deeply into the meaning of the inverse relationship between payout and investment performance. Why should payout not be higher in foundations with higher rates of return? What does the failure of payout performance to march in lockstep with investment performance tell us about foundation payout policy and practice? And, what does this tell us about the impact of the 1981 change in the payout requirement? We turn to these questions in the next section.

II.
EXPLAINING FOUNDATION PAYOUT PERFORMANCE

Aside from the vagaries of good or bad fortune, two broad sets of factors affect the ratio between foundation payout, i.e., their grants and program expenditures, and foundation assets: payout policy and investment strategy. The first of these shapes payout directly. The latter does so indirectly, by affecting the form of investment return—annual income or appreciation. These two sets of factors are closely interrelated, but they require separate analysis.

Payout Policy and Payout Performance

An important part of the explanation for the patterns of payout performance outlined above may be that foundations simply follow different policies in setting their payout levels. The crucial issue here is whether payout policy determines investment policy or investment policy calls the tune for the grants policy.

For example, payout decisions can be made independently of investment performance on the basis of perceptions of programmatic need. This was the course that the Ford Foundation apparently pursued during the 1960s when it set spending levels well in excess of investment returns and cut significantly into its asset base.

Alternatively, foundations can treat payout as essentially an aspect of investment management and arrange their investments in such a way as to avoid paying out more than the minimum required by law. Such a course is most likely to promote the long-term growth of the foundation's asset base but at the expense of the current level of payout.

Finally, foundations interested in high payout can pursue a relatively conservative investment policy, investing heavily in fixed-income assets and then paying out all the income from such assets. Such an approach is likely to support a payout rate above the 5-percent minimum, but it will tend to limit the long-term growth of the foundation's investment portfolio and, potentially, the absolute size of qualifying distributions over the long term.

Table 5.4
Foundation Payout Policy by Size of Foundation

Payout Policy	Percent of Foundations			
	All (N = 8,684)	Large[a] (n = 44)	Medium[a] (n = 70)	Small[a] (n = 108)
Pay out all investment income	37.5	9.1	8.6	37.0
Payout shaped by investment performance	30.8	6.8	31.4	28.7
Pay out what law requires	14.1	36.4	28.6	16.7
Payout shaped by program goals	7.4	13.6	8.6	5.6
Payout determined by gifts	4.8	—	—	1.9
Pay out slightly more than required	2.9	34.1	21.4	1.9
Other	2.5	—	1.4	8.2
Total	100.0	100.0	100.0	100.0

Source: Salamon/Voytek Foundation Investment Study.
[a] See Table 5.2.

Our survey responses reveal that such variations in payout policy are clearly evident among foundations. Table 5.4 shows that the most common approach, reported by close to 40 percent of the foundations, was to set payout equal to investment income. Another 30 percent of the foundations reported that payout decisions are shaped by overall investment performance, including appreciation, rather than merely by the amount of investment income. A much smaller 14 percent of the foundations reported paying out only what the law requires. The most striking finding is that a mere 7 percent indicated that the primary determinant of payout levels is programmatic goals.

Here, also, the proportions of foundations that fell into these different categories varied considerably by the size of the foundation, as Table 5.4 also makes clear. For example, only 14 percent of all foundations reported that they pay out only the minimum that the law requires and another 3 percent that they pay out just slightly more than the law requires, but about 70 percent of the largest foundations reported one of these two responses. In contrast, among the small foundations, investment performance (particularly investment income) plays the dominant role in determining payout levels.

What this suggests is that at least two very different approaches to payout decision making are evident among foundations and that size is an important factor in explaining how foundations are distributed between them. In particular, among the smaller foundations investment decisions are most likely to determine payout levels. In contrast, among the larger foundations payout considerations seem more likely to shape investment decisions. Put somewhat differently, the small foundations tend to pursue a conservative investment approach fo-

cusing on fixed-income securities and then, whether by choice or necessity, pay out all the earnings this investment approach generates. In contrast, the larger foundations are more likely to devise an investment strategy that takes account of the payout requirement and to arrange their investments in such a way as to minimize the impact of the payout requirement on the long-term asset base of the foundation.

This finding is consistent with the earlier finding that larger foundations tend to have more formalized and coherent investment strategies. What becomes clear is that these formalized investment strategies embrace payout decision making as well: Investment decisions are made with the payout requirement in mind. This may help to explain why the payout rates of the larger foundations are generally lower than those of the small foundations.

One additional explanation for these observed differences may be that the smaller foundations are more likely to be dominated by the original donors or their immediate families and to be more interested in giving money away in the foreseeable future rather than holding it over the longer run. This would lead them to opt for low-risk/high current yield investments, with an accompanying decision more or less to give away whatever is earned. The larger foundations, run by professional staffs, are likely to have a stronger commitment to operating in perpetuity and therefore to selecting investment and payout policies that encourage this.

That these payout *policies* actually affect payout *rates* is evident in Table 5.5. What this table shows is that foundations reporting that payout levels are shaped by investment performance generally had higher payout levels than foundations reporting a policy of paying out the minimum required by law. Over 20 percent of the first group had payout rates exceeding 10 percent of assets, compared to only 6 percent of the second group.

Table 5.5
Relationship Between Payout Rate and Payout Policy

Payout Policy	Percent of Foundations with Payout That Is:			
	5 Percent or Less (*n* = 32)	5.1–9.9 Percent (*n* = 148)	10 Percent+ (*n* = 42)	Total
Pay out what law requires	29.6	64.8	5.6	100.0
Pay out slightly more than law requires	21.9	65.6	12.5	100.0
Payout shaped by investment performance	7.2	71.4	21.4	100.0
Payout shaped by program	11.1	61.1	27.8	100.0
Pay out all investment income	4.0	70.0	26.0	100.0

Source: Salamon/Voytek Foundation Investment Study.

Investment Strategy and Payout Rates

That this relationship between payout policy and payout performance holds is probably due, however, to variations in the investment strategies that foundations pursue, and in particular to the kinds of assets they hold. As Chapter Three makes clear, foundations vary considerably in the asset mixes that they select, and these variations are related to the investment goals the foundations are pursuing. Foundations pursuing a total-return objective are likely to have a larger proportion of their assets in common stock, whereas those seeking only to maximize current income may be more heavily invested in fixed-income securities.

Chapter Four confirmed that foundations heavily invested in common stock achieved a higher total return but a lower current-income yield than foundations heavily invested in fixed-income securities during 1979–1983. On the other hand, such foundations also tend to be exposed to higher levels of risk and variability in their current returns.

These prior findings help to explain the pattern of payout performance reported in this chapter. Because of the high interest rates prevailing during the 1979–1983 period, the 1969 requirement that foundations pay out the higher of their total annual investment earnings or 5 percent of their asset value meant that foundations heavily invested in fixed-income securities

could be expected to have the highest payout rates because their income yield would exceed 5 percent of their assets. In other words, a generally conservative investment strategy featuring heavy reliance on fixed-income securities instead of common stocks could be expected to produce a higher-than-average payout rate.[4]

Table 5.6 confirms this, showing that the greater the share of assets in common stock, the lower the payout rate. Thus, foundations with low proportions (less than 40 percent) of their assets in common stock registered a median payout rate of 8.6 percent, whereas those with most (60 percent or more) of their assets in common stock registered a median payout rate of only 6.2 percent.

Table 5.6
Foundation Asset Composition and Payout Rates, 1979–1983

Percent of Assets in Stock	Median Annual Payout Rate, 1979–1983 (Percent)
Less than 40	8.6
40–50.9	6.4
60 or more	6.2

Source: Salamon/Voytek Foundation Investment Study.

Although the quality of the asset composition data makes it appropriate to interpret these findings with some caution, there is nevertheless a reasonable basis here for explaining why the smaller foundations tend to have higher payout rates than the larger ones. The smaller foundations tend to have an investment strategy that favors fixed-income securities, a shorter time frame, and less risk. These foundations achieved lower total-return rates but higher income yields. With a median income yield well in excess of 7 percent, the small foundations naturally tended to have payout rates in excess of 7 percent, since this is what the law required. In contrast, the larger foundations tended to pursue a different strategy and earned larger total returns but lower income yields. Since the income yield of the larger foundations was lower, their payout rate consequently could be lower as well.

[4] It is important to recall here that payout *rate* is not the same as payout *amount*. Payout rate is the payout amount *divided by* the average annual fair market asset value. If the asset value grows slowly, or declines in value, therefore, the payout rate can increase even though the payout amount holds steady or declines. A low payout rate applied against a rapidly growing asset base can thus produce more grants than a high payout rate applied against a constant or declining asset base.

III.
IMPACT OF THE 1981 PAYOUT REQUIREMENT CHANGE

These findings, in turn, have implications for the likely impact on payout performance of the change in the payout requirements enacted in 1981. This change was designed to give foundations greater freedom to adjust their investment policies to market conditions by eliminating the requirement that they must pay out all of their investment earnings. In this section, we examine what impact this change has had, or is likely to have, on payout behavior.

Impact on Payout Policy

A natural starting point for this analysis is the impact of the payout requirement change on the payout *policies* of foundations. About 15 percent of the foundations represented in our sample reported that the procedure they used for determining their payout rate had changed since 1981, and the vast majority of these credited the payout requirement revision with a significant share of the responsibility for the shift. The most frequently cited change was away from setting payout at what the law required and toward having payout shaped by investment income or overall investment performance (see Table 5.7). In other words, foundations expressed greater freedom to adjust payout rates to investment conditions and performance, as advocates of the change in the law sought.

Table 5.7
Changes in Foundation Payout Policies Since 1982

	Percent of Foundations	
Payout Policy	Pre-1982 Policies of Foundations Making a Change (N = 3,371)	Post-1982 Policies of all Foundations (N = 22,476)
Pay out all investment income	32.5	37.5
Payout shaped by investment performance	11.8	30.8
Pay out what law requires	41.2	14.1
Payout shaped by program needs	5.7	7.4
Payout determined by gifts	5.7	4.8
Pay out slightly more than required	3.1	2.5
Total	100.0	100.0

Source: Salamon/Voytek Foundation Investment Study.

Impact on Payout Rates

Has this greater freedom to set payout policy led to a significant decline in payout rates, as some opponents of this liberalization of the payout requirement feared? The evidence available to date suggests that the answer to this question is no. Table 5.8 shows that the average annual median payout rate for the foundations we examined declined only slightly between the period just prior to the change in the payout law and the period just after it. The median payout rate stood at 7.7 percent a year during 1979–1981 and at 7.5 percent a year during 1982–1983.

To the extent that any change did occur, moreover, it took place among the smaller foundations. In contrast, the payout rate for the large- and medium-sized foundations held steady between these two periods.

Table 5.8
Foundation Payout Rates, 1979–1981 versus 1982–1983

	Percent	
Size Class of Foundation	1979–1981	1982–1983
Large (n = 88)[a]	6.0	6.0
Medium (n = 252)[a]	6.4	6.4
Small (n = 119)[a]	7.7	7.5
All (median) (N = 22,476)	7.7	7.5

Source: Salamon/Voytek Foundation Investment Study.
[a] See Table 5.2.

One reason for this lack of change in the payout rate before and after the liberalization of the payout law may be that 1983 was too soon for any change to show up. To check on this, we examined the payout performance of 75 foundations with fiscal years ending in December for which we were able to compile complete investment and payout data for 1984 as well. The results of this analysis, reported in Table 5.9, indicate that a more noticeable decline in payout rate did occur in 1984 for these foundations. In particular, the median payout rate for these 75 foundations fell from 6.2 percent in 1982–1983 to 5.7 percent in 1984.

Table 5.9
Payout Performance of 75 December Fiscal Year Foundations for Which 1984 Data Are Available

	Percent	
Time Period	Median-payout Rate	Median-income Yield
1979–1981	6.2	7.1
1982–1983	6.2	7.7
1984	5.7	6.6

Source: Salamon/Voytek Foundation Investment Study.

However, market conditions also changed in 1984. In particular, yields on government bonds declined. Table 5.9 shows that this was reflected in declines in the income yield of these same foundations. In fact, the drop in income yields was much sharper than the drop in the payout rate. Thus, there is probably as much reason to attribute the decline in the payout rate in 1984 to the decline in current income yields as to the change in the payout requirement. Even in the absence of the change in the payout requirement, in other words, the dip in the payout rate would probably have occurred anyway.[5]

How can we explain this situation? Why has the change in the payout requirement not had more of an impact on foundation payout performance? One important clue to the answer to this question is probably embedded in the analysis of payout patterns covered in the previous section. A central conclusion of that analysis was that the relatively high median foundation payout rate over the 1979–1983 period resulted largely from the performance of the smaller, less aggressive foundations, which tend to invest more heavily in fixed-income securities and to pay out all of the resulting investment earnings.

This has two important implications for the impact of the 1981 change in the payout requirement. In the first place, it means that the major, direct beneficiaries of the 1981 payout requirement change are paradoxically likely to be the smaller foundations with the least active investment approaches and the lowest rates of total return. These are the foundations that had the highest payout rates under the old law, at least in part because of the requirement that they pay out all of their investment earnings. Therefore, they are the ones that have the greatest room to lower their payout rates to the 5-percent minimum. In contrast, the larger foundations had already adjusted their portfolios in such a way as to minimize their payout rate and preserve their asset base. While the payout requirement change might be expected to alter the *investment* behavior of these foundations, it is not likely to alter their payout behavior.

But if the smaller, more risk-averse foundations stood to gain the most from the change in the payout requirement in 1981, they were also the least likely to take advantage of the new opportunities. These were, after all, the foundations whose payout performance depended most heavily on their investment performance and whose investment performance in turn was shaped by the most conservative preferences. Payout requirement or not, these foundations could be expected to place the bulk of their assets in relatively dependable, fixed-income securities and then dedicate the annual proceeds of these investments to their charitable activities. Since these are the foundations that had the most room to respond to the change in the payout requirement, it becomes clear why the payout requirement change has had less impact on payout rates than might have been feared.

Payout Variability

One final potential impact of the change in the payout requirement concerns not the overall scale of the payout level but its variability. One of the major complaints about the 1969 payout requirement was that it exposed foundation grant making to excessive year-to-year fluctuations by tying the minimum distribution requirements to fluctuations in the market value of assets. This criticism was especially applicable during the 1970s because of unusually wide fluctuations in interest rates and dividend yields. For example, between 1973 and 1982, the Bush Foundation in St. Paul, Minnesota, experienced year-to-year fluctuations in its required "qualifying distributions" of as much as 50 percent. In seven of these nine years, moreover, the portion of this fluctuation attributable to the investment income feature of the payout requirement, which was eliminated in 1981, exceeded the portion attributable to the asset value feature of the requirement. Had the 1981 change in the payout requirement been in force during this time period, therefore, the payout rate could have been much less volatile and foundation grant planning considerably improved. Although the carryover provisions of the payout law provided some relief, it was only partial. Humphrey Doermann, president of the Bush Foundation, explains:

> For us, at least, it often takes two years from the time of a seemingly bright idea until the time it is recognized to be feasible and the first program grants are awarded. Once started, in turn, our best programs have tended to last for five to seven years. My observation of the 1970s was that frequent and sizable changes in private foundation required distribution rates inhibited optimum grantmaking performance in foundations like ours because it discouraged careful planning. . . .[6]

By eliminating one important source of year-to-year fluctuation in the required payout rate, the 1981 change therefore probably contributed to effective grant making quite apart from any impact it had on payout rates.

[5] These results do not mean that the absolute level of grants necessarily declined in 1984 because equity asset values had increased sharply during 1983.

[6] Letter from Mr. Humphrey Doermann, president of the Bush Foundation, to Elizabeth Boris, vice president for research of the Council on Foundations, June 3, 1987.

IV.
CONCLUSION

Private foundation program expenditures have thus run at a relatively high ratio to foundation assets during the period covered by this study. In fact, the average median payout rate over the period 1979–1983 was half again as great as the 5-percent minimum required by the 1969 tax law and retained in the revision to that law enacted in 1981.

However, this relatively high payout rate for foundations as a group resulted in large part from the relatively conservative investment strategies of a large number of generally smaller foundations. Heavily invested in fixed-income securities, these foundations benefited from the unusually high return rates on such securities during this time period and boosted their payout rates accordingly. Because this investment strategy yielded less growth in the asset base, however, the growth in the actual amount of qualifying distributions barely kept pace with inflation even though the median payout rate was high.

The 1981 change in the payout rate may alter this situation by allowing foundations to retain investment earnings in excess of 5 percent of their assets. To the extent it does so, however, the major direct beneficiaries, paradoxically enough, are likely to be the foundations with the more conservative investment strategies, as these are the foundations that have had payout rates most in excess of 5 percent. But these are also the foundations least likely to alter their investment policies in response to the change.

The 1981 payout change therefore increases the importance of improving the investment performance and investment management of these mostly smaller foundations. Since these foundations are now able to plow more of their earnings back into investment, it is crucial that they do so wisely when they opt for reinvestment. Otherwise, the liberalization of the payout requirement may not produce the gain in foundation assets that its advocates hoped for.

For the larger and more aggressive foundations, the major impact of the 1981 change is likely to occur on the investment, rather than the payout, side of the ledger. These foundations have already made adjustments in their investment behavior to minimize the impact of the payout requirement on their asset base. Therefore, it is unlikely that the liberalization of the payout requirement will reduce their payout rates very much more, if at all. The more likely consequence is to free these foundations to take more advantage of high-yield investments when market conditions permit. It may allow them to achieve higher returns with less short-term risk. Beyond this, the change seems likely to reduce somewhat the year-to-year fluctuations in required distributions and, thus, to contribute to improved grant making and program development.

Chapter Six

CONCLUSIONS AND RECOMMENDATIONS

Foundations, as we noted earlier, are not simply investment pools dedicated to profit maximization or long-term asset growth. By law, foundations have a more complex set of objectives and a more confining set of constraints. The principal legal rationale for the existence of foundations is to foster charitable activity through the distribution of grants or other services. This means that foundations are only incidentally in the business of accumulation; they exist to be in the business of distribution.

Nevertheless, the asset base of foundations is essentially their only source of income to finance grants. Foundations must therefore manage their assets in such a way as to generate the stream of income they need to support their charitable expenditures. In the process, they must also balance the needs of current claimants against those of future claimants, which means that they must maintain, or increase, the real value of their assets over time while generating a reasonable level of annual grant support.

Since 1969 foundations have had to pursue these competing objectives within a constraining regulatory structure that has stipulated minimum payout requirements. Meeting these payout requirements has not only limited the amounts to be set aside to provide for higher grants in the future. In an inflationary environment, high payouts have led to persistent inroads into the real purchasing power of foundation capital. Finally, the regulations may have encouraged some foundations toward high-yield investments because of the risks involved in paying out 5 percent or more of a portfolio heavily committed to common stocks. In the long run, such a policy can lead to lower rates of growth in foundation capital and, therefore, to a reduction in the ability of foundations to sustain the real value of their grants in the future.

I.
THE RECORD

Against the backdrop of these constraints, the data presented in this report suggest a reasonable, if mixed, record of foundation investment and payout performance. The overall annual rate of return on foundation assets for the period we examined, 1979–1983, exceeded the comparable rate for a control portfolio containing a conventional asset mix. What is more, we found solid evidence that a significant portion of the foundation community has adopted key features of modern investment management, emphasizing total-return objectives, balancing of risk against return goal and time frame, diversification of assets, and use of competitive, professional management. Finally, payout performance was also strong. The median, or foundation-weighted, payout rate exceeded the legal minimum by 50 percent over the period we examined, and the asset-weighted rate exceeded the minimum by 20 percent.

Nevertheless, although there were significant signs of strength in the record of foundation investment and payout behavior that we examined, there were also significant areas of weakness. While the overall rate of return on foundation assets exceeded the market averages, this was primarily because of the performance of the relative handful of larger foundations. In contrast, most foundations performed below the control portfolio.

In fact, the rate of return the median foundation achieved was not sufficient to support a minimum 5-percent payout rate and still preserve the real, inflation-adjusted value of the asset base. When account is taken of the fact that the median foundation had an average payout rate closer to 7.5 percent than to 5 percent, it becomes clear that this was a period of significant decline in the real value of the assets of most foundations, despite the overall growth that occurred.

In addition, only a fraction of the foundation universe made use of an active investment management approach. For most foundations, explicit investment objectives and strategies were rare, board involvement in the investment function limited, return goals and risk tolerance restricted, the time frame over which results were measured rather narrow, and the overall investment management style hands-off rather than hands-on. Investment management for a significant portion of the foundations consisted of turning the assets over to outside managers to be invested in low-risk, fixed-income securities.

To be sure, there is nothing inherently wrong with this style of investment management, but at least during the period we examined there is evidence that it exacted a price in terms of lower returns. What is more, we believe that this tendency toward lower returns was not peculiar to this period. In fact, the period we examined probably *understated* the relative superiority of the more active approach, since the unusually high rates of current return on fixed-income securities and cash equivalents during the 1979–1983 period probably gave a greater boost to the more conservative institutions than they would have had at most other times.

Therefore, despite the progress that has apparently been made in improving foundation investment and payout management over the past decade or two, much remains to be done. This is particularly true in view of the fact that the institutions that performed worst during the 1979–1983 period are the ones that stand to gain the most from the 1981 payout requirement change.

What further steps can be taken, then, to improve foundation investment and payout processes, and foundation investment performance?

Full answers to this question would take us well beyond the limits of this report. Nevertheless, it is possible to identify at least briefly some of the avenues of change that seem most promising. While none of these changes can cancel out the impact of market develop-

ments over which foundations have little control, they can at least help foundations cope with these developments more effectively. Four such types of changes seem worth noting.

II.
FOUR MODEST PROPOSALS

Extend the 1981 Payout Change

When Congress liberalized the payout requirement in 1981, it warned that the change could be rescinded if foundations abused it. A first issue that must be faced with regard to foundation investment and payout policies is thus whether the payout change enacted in 1981 should be extended.

Although the data available to date do not make a definitive answer to this question possible, the preponderance of the evidence certainly supports an extension. At a minimum, there is little evidence that the liberalization of the payout requirement by itself has led to any significant reduction in the payout rate of foundations, and there is strong reason to doubt that it will. For one thing, as we have seen, the larger foundations seem to have chosen investment strategies that have already left them with payout levels close to the 5-percent minimum. While the smaller foundations tend to have higher payout rates, this is largely the product of investment decisions that seem to be made independently of the payout requirement. While the payout rates of these foundations could decline, this is likely to happen more as a result of changes in income on fixed-income securities than as a result of changes in the payout law.

More positively, while not risking great loss in payout levels, the liberalization of the payout requirement does seem to have freed the more active foundations to pursue a more balanced investment approach. It has done so by allowing foundations to retain some of the income from fixed-income securities when interest rates rise, thus avoiding discouragement of such investments. With an increase in interest rates at least a reasonable prospect in view of existing trade and budget deficits, this greater freedom seems desirable to maintain.

Multiyear Basis for Payout Computation

While the 1981 change in the payout requirement was a step in the right direction that should be extended, it may not have gone far enough. A useful next step would be to lengthen the time frame that is used to compute the minimum payout requirement that still remains. Under the current law, foundations must pay out in grants and other "qualifying distributions" in any given year 5 percent of the average market value of their assets during the year.[1] Because asset values can vary widely from year to year because of market swings, however, this produces sizable shifts in required distribution amounts from year to year.[2]

Although the "carry-over" provisions in the payout law can reduce these fluctuations, they can do so only partially.[3] Decoupling payout rates from investment income, as was done in 1981, will also help, as noted in Chapter Four. But this will not affect the fluctuation that results from changes in the market value of assets, as opposed to changes in dividend and interest earnings. One way to deal with this remaining fluctuation is to base the payout requirement not on a single year's asset value, but on a floating average of the asset value over the prior *three (or five) years.*

Two major findings of this research argue for this change. The first is the discovery that the best predictor of return rate was risk tolerance—the willingness to take some risks. Generally speaking, foundations that were willing to take reasonable risks achieved higher rates of return. But willingness to take risks is importantly related to the time frame over which foundations evaluate results. Generally speaking, the shorter the time frame is over which results must be demonstrated, the lower is the level of risk one can assume, and the more cautious one must become. The existing payout requirement, by tying payout rates to the asset value in a given year, has the effect of shortening the time frame over which foundations can manage their investments, thus lowering the risk they can tolerate. It thus works counter to effective investment management.[4]

---

[1] Formally, foundations have until the end of the subsequent fiscal year to make up any shortfall in payment in a given year.

[2] For example, the Standard and Poor's 500 stock index generated a return of 32 percent in 1980, −4.9 percent in 1981, and 21.4 percent in 1982.

[3] Foundations are permitted to "carry forward" excess payout for up to five subsequent years, reducing the amount that must be paid out of the current year's income. However, this only affects foundations with "excess payout."

[4] Robert E. Swaney, Jr., vice president for investments at the Charles Stewart Mott Foundation, has made this point convincingly: "I believe that the most significant factor that causes foundations to want to minimize market volatility in their portfolios is the requirement that payout must be based on a one year average portfolio value and paid out in the subsequent year. It is likely that if the payout were based on a five year average value (or even longer) moving forward one year at a time that foundations would have a more stable and predictable payout pattern. This stability could then appropriately lead to the assumption of greater portfolio volatility or risk and to higher long term investment returns." Robert E. Swaney, Jr., to Elizabeth Boris, July 8, 1987.

A second argument for lengthening the time frame over which the required payout level is computed flows from the observation that high levels of variability in payout levels can negatively affect the foundation grant-making function. With required payout levels that increase or decrease by 50 or more percent a year as a result of market fluctuations, foundations sometimes find themselves required to distribute more money in a given time frame than they consider prudent or to cut heavily into programs to which they have already made multiyear commitments.

A payout rate based on a three-year floating average of asset values would significantly reduce these problems. Foundations could then take a somewhat longer perspective on both their investment and grant programs, and both the overall rate of return and the scale and effectiveness of grant activity could benefit.

More Attention to Foundation Investment Management

While our data indicated that a number of foundations have developed fairly sophisticated approaches to the management of their investments, they also indicated that the majority clearly have not. Indeed, there is some indication that a large number of foundations are not even aware of the options.

What this suggests is the need to focus more attention on the investment function of foundations as opposed to their grant-making function. This could include special training sessions, more attention to investment matters at the annual meetings of the Council on Foundations or of regional associations of grant makers, and printed materials disributed through *Foundation News* or other trade outlets. While there are numerous commercial businesses eager to sell investment services of various sorts to foundations, there is too little effort put into helping foundations sort out what kinds of help they need and how their investment management process could best be structured.

Attention to foundation investment management does not only involve instructions for foundations, however; it also involves keeping much better track of how foundations perform. The absence of a regular database on foundation investment activities is symptomatic of the general neglect of this facet of foundation operations and a cause for great concern. While the work described in this report provides the beginnings of such a database, a more permanent process of data gathering and analysis will be needed if investment management is to be given the kind of focus and priority it deserves.

A Common Fund for Foundations

Finally, the foundation world could usefully borrow a page from the education field in deciding how to handle the challenges of investment management. Concerned about their inability to afford professional investment management and keep pace with rapidly changing market developments, a group of 62 colleges and universities banded together in 1971, with help from the Ford Foundation, to create the Common Fund, a nonprofit organization charged with the responsibility of overseeing the management of all or parts of these schools' investment assets. By the mid-1980s, the Common Fund included over 750 colleges, universities, and independent schools and was responsible for the management of more than $2.4 billion in assets.

By pooling resources, the Common Fund is able to achieve economies of scale in management costs and attract better investment management talent than would likely be available to its members on their own. It does so, moreover, while respecting the objectives of its member institutions by offering a variety of investment funds that provide different mixes of risk and return. While none of this is a guarantee of high return, the Common Fund's record is impressive, both with respect to management costs and with respect to investment performance. As a recent *Forbes* article put it: "The fund has almost certainly done better than many of its member schools would have been able to do on their own."[5]

Given the evidence presented in this report, it seems clear that something like the Common Fund might be a welcome addition to the foundation field as well. This is particularly true in view of the fact that the vast majority of foundations are so small. If these foundations are to benefit from professional management and take full advantage of the new freedom that the relaxed payout requirement now gives them, they would be well advised to consider pooling at least a portion of their resources and creating an investment management instrumentality that can better serve their unique investment needs. Even large- and medium-sized foundations could benefit from such an approach since, despite their size relative to other foundations, they are frequently rather small players in the larger world of institutional investment. This has been the experience, at any rate, with the Common Fund for education, which has attracted large, as well as small, institutions. A Foundation Common Fund, or access to the existing Common Fund pools, might be a useful way to boost the return rates of foundations and, hence, their payout levels, while protecting the existing institutional structure of the foundation world. It is certainly an idea worth exploring.

[5] Richard Phalon, "Shotgun Approach," *Forbes* (June 15, 1987), p. 134.

III.
THE TASK AHEAD

In the wake of government budget cuts and continued constraints on government spending, private foundations have gained renewed prominence in American life. As public funding has declined, nonprofit organizations have turned increasingly to private foundations for the support they need to survive.

The ability of foundations to respond to these demands and contribute meaningfully to the solution of public problems is critically affected, however, by the success with which they manage the assets in their control and balance the competing demands of current recipients and future claimants. Because of difficulties in the past, foundations in 1969 lost a considerable portion of the freedom they formerly had to manage their fiscal affairs on their own. In the process, foundations were subjected to a regulatory regime that had at least some counterproductive consequences.

Beginning in 1981, Congress moved to liberalize these restrictions by permitting foundations to reinvest at least a portion of their investment earnings. The evidence reviewed here generally confirms the wisdom of this move and suggests the desirability of an extension of this liberalization and possibly some further change in the same direction.

To the extent such change occurs, however, foundations must recognize that greater freedom brings with it greater responsibility. Greater responsibility is needed, moreover, not only to avoid the abuses that led to the restrictions in the first place, but also to help achieve the kind of substantial long-term growth that the foundation community and those that rely on it for support require.

As we have seen, substantial numbers of foundations have developed investment approaches that seem likely to achieve this result. But far more have not, and the philanthropic community in general has suffered. In the face of significant new opportunities and demands, this situation is highly undesirable. It is therefore important to continue the improvement in foundation investment behavior that is already underway and extend it to the considerable number of foundations that still seem to be lagging behind. By demonstrating that at least some foundations are performing quite well while others still have some distance to go, we hope that this report will provide both a basis for protecting the gains that have been made and a stimulus for encouraging the changes that are still needed.

Foundation investment management is too important to the fiscal health of the nonprofit sector and, through it, to the social and political health of the nation to remain hidden from view. It must now be given the greater attention it deserves.

Appendix A

METHODOLOGY

Two basic sources of data were utilized in this project: first, a mail survey distributed to a sample of private foundations throughout the country, and second, a body of financial data drawn from the 990-PF forms that foundations file annually with the Internal Revenue Service. Slightly different samples ultimately were utilized for these two data sources. In addition, because the distribution of foundations differs markedly from the distribution of foundation assets, different weighting systems had to be used to blow up the respective samples to the universe of foundations and the universe of foundation assets. This appendix discusses these two data sources, the samples used for each, the weighting systems that were used, and the "confidence intervals" associated with each.

THE MAIL SURVEY OF FOUNDATIONS

To gather information on foundation investment and payout policies and processes, a mail survey was distributed to a sample of private foundations. All private foundations except for community foundations and corporate foundations were covered in this survey. The discussion below reviews the survey sample and the survey process.

Survey Sample

The survey sample was generated from computer tapes supplied by the Foundation Center. These tapes included the names and addresses of all foundations and, in the case of foundations with assets in excess of

$1 million, designated the type of foundation (nonoperating, community, operating, and corporate) and the asset size.

Company-sponsored and community foundations were excluded from the files from which the final sample was drawn. This left a total of 22,294 foundations.

Because of the maldistribution of foundation assets (the top 0.7 percent of all foundations account for 60 percent of all foundation assets) a disproportionate stratified random sampling method was used. Such an approach was designed to make sure there would be a sufficient number of large foundations in the final sample to support inferences about foundations in this size class and, hence, about the majority of foundation assets.

In particular, we subdivided the list of foundations into four size classes on the basis of the foundation's assets as of 1982:

$50 million or above:	166 foundations
$10–49.9 million:	533 foundations
$1–9.9 million:	2,475 foundations
Less than $1 million:	19,120 foundations

Separate random samples were then taken from each of these four size classes. These samples were "disproportionate" to ensure adequate coverage of the size classes with the fewest foundations. What is more, account was taken of the past record of foundation responses to surveys of this sort. That experience suggested that foundations do not typically respond to surveys and that the response rate is lowest among the

Table A-1
Sample Sizes by Data Source

Size Category	Survey		990-PF, 1979–1983		Survey plus 1979–1983 990s	
	Number	Percent	Number	Percent	Number	Percent
$50 million and over	74	15	88	19	45	22
$10–49.9 million	125	26	252	55	77	38
Below $10 million	280	59	119	26	83	41
Total	478	100	459	100	205	100

smaller foundations. This in turn suggested the need to have a large initial sample so that the final returns would be adequate to support the kind of analysis that was desired. Thus, a 100-percent sample was taken of foundations in the two largest size classes; a 30-percent sample was taken of foundations in the $1–9.9 million size class, and a 20-percent sample was taken of foundations in the smallest size class. This yielded a potential sample of 5,348 foundations, from which we hoped to get approximately 1,000 responses.

Survey Development and Execution

The survey instrument was developed with the assistance of a project advisory committee composed of leading experts on foundation fiscal management. A draft of this instrument was reviewed by this committee in December 1984 and then field-tested on 30 randomly selected foundations in the spring of 1985. Following the field test, the survey form was revised, reviewed again by the advisory committee, and then distributed to our 5,348 potential respondents in October 1985. A reminder postcard was then distributed two weeks later and a second survey form sent to nonrespondents in mid-November. In addition, advisory committee members were enlisted to encourage responses from the larger foundations that had not submitted completed forms by early 1986.

Altogether, 527 foundations completed survey forms in whole or in part, as noted in Table 1.1 in the text. Of these, 49 turned out to be community foundations, leaving 478 valid independent foundations.

Completed survey forms were subjected to an in-depth editing process, and scales were developed for open-ended responses. The data were then keypunched and prepared for computer analysis.

FORM 990-PF DATA

Data from 990-PF forms filed by foundations were gathered from microfiche files available at the Foundation Center. Information was sought for the six years between 1979 and 1984 on all foundations with assets in excess of $10 million and on all those foundations with assets below $10 million that also responded to

our survey (a total of 1,003 foundations). Twenty-three data elements were collected from the 990-PF forms for each foundation for each year covered.

Because of gaps in data availability, complete data files even on the 1979–1983 time period could be assembled on only 459 of the 1,003 foundations targeted for coverage. For the longer time period, 1979–1984, complete records were available on only 184 foundations. Both survey and 990-PF data for 1979–1983 were available on 205 of the foundations. Table A-1 summarizes this data availability, showing the total number of foundations covered by each data source or combination of data sources and the breakdown for each source among foundations of different sizes. Significantly, despite the gaps in coverage and problems of data availability, our research design yielded enough responses in each size class to sustain statistically meaningful observations even for the smallest of the subsamples, i.e., the set of foundations for which we had both survey and 1979–1983 990-PF data.

WEIGHTING SYSTEMS

In order to go from the survey responses and the 990-PF data to the universe of foundations, two different weighting schemes were developed. The first applied mostly to the survey data and sought to adjust the data to reflect the full universe of foundations. The second applied mostly to the 990-PF data and sought to adjust these data to reflect the universe of foundation *assets*.

Weighting by Number of Foundations

To develop the foundation-weighted results, the number of survey respondents in each size class was divided by the number of foundations of that size class in the full universe. The resulting "weighting factor" was then applied to each response of each foundation in that size class. For example, the survey sample included 74 foundations with assets of over $50 million, compared to 166 in the full population. Each large foundation was therefore assigned a weight of 2.24 in calculating the overall results. Table A-2 reports the foundation-based weights for each of the three size classes of foundations into which we broke the data.

Table A-2
Weighting Factors, Foundation-weighted Survey Results

Size Class	Foundations in		Weighting Factor (1) ÷ (2)
	Universe (1)	Sample (2)	
Large ($50 million +)	166	74	2.24
Medium ($10–49.9 million)	533	124	4.26
Small (Under $10 million)	21,595	280	77.40
Total	22,294	478	

Because the 990-PF sample differed slightly from the survey sample, different weighting factors had to be used to generate foundation-weighted results from the 990-PF data set. Table A-3 shows the derivation of these weighting factors.

Table A-3
Weighting Factors, Foundation-weighted Rate-of-Return and Payout Results

Size Class	Foundations in		Weighting Factor (1) ÷ (2)
	Universe (1)	Sample (2)	
Large (50 million +)	166	88	1.89
Medium ($10–49.9 million)	533	252	2.16
Small (under $10 million)	21,595	119	181.50
Total	22,294	459	

Weighting by Assets

A similar method was also used to go from the 990-PF financial data to the full universe of foundation *assets*. Because the distribution of assets differs from the distribution of foundations, and because we had different proportions of large, medium, and small foundations in the 990-PF subsample, different weightings had to be developed for this purpose. Table A-4 records these weighting factors and the basis on which they were computed.

To clarify when weightings have been used, we have recorded sample sizes (designated by lower case *n*) in the tables where just the sample results are being recorded (typically the results by size class), and the universe size (shown by capital *N*) where the weighted results are being recorded. Because of missing responses on some particular questions, the universe *N* may differ from table to table.

STANDARD ERROR OF THE ESTIMATES

Any sampling approach involves the possibility that the results reported for the sample deviate from the true characteristics of the full population from which the sample is drawn. The likelihood that this is the case varies with two things: (a) the size of the sample, and (b) the observed distribution of results in the sample. As far as the first of these is concerned, a sample size of 50 is generally regarded as the minimum necessary. As far as the latter is concerned, the closer to a 50–50 split exists in the sample, the greater is the likelihood that the sample result and the population result will diverge.

Table A-4
Weighting Factors, Asset-weighted Rate-of-Return and Payout Calculations

Size Class	Foundation Assets (in billions of Dollars)		Weighting Factor (1) ÷ (2)
	Universe (1)	Sample (2)	
Large ($50 million +)	25.4	14.7	1.72
Medium ($10–49.9 million)	9.7	4.0	2.42
Small (under $10 million)	12.4	0.2	57.40
Total	47.5	18.9	

To express the degree of confidence one can have in the results of an analysis resting on a sample, statisticians compute a "standard error of the estimate" and a "confidence interval" for the results. The confidence interval expresses the range on either side of an observed sample result that one can expect the true value for the full population to fall. The greater the degree of confidence one wants to have, obviously, the wider the confidence interval will be. By convention, statisticians use a 95-percent confidence interval. This means that one is 95 percent confident that the true population proportion is within the specified range of the proportion reported for the sample.

The formula for computing the 95-percent confidence interval for a sample like ours is:[1]

$$I_c = \sqrt{\frac{Pq}{n} \times \frac{N-n}{N-1}} \times 1.96$$

Where:
- I_c = confidence interval
- P = sample proportion
- $q = 1 - p$
- n = sample size
- N = population size

Using this formula, Table A-5 reports the confidence intervals for the full survey sample and for the three subsamples. What this table shows is that for the full sample, we are 95 percent confident that the true population proportion ranges from between ± 1.0 percent to ± 4.5 percent of the observed proportion. For the large foundation subsample, the confidence interval ranges from ± 1.7 percent to ± 8.6 percent. This means, for example, that if we find 50 percent of the foundations in our sample have a certain characteristic, we are 95 percent certain that the true value in the universe of foundations falls between 45.5 percent and 54.5 percent.

Table A-5
Standard Errors of Estimates for Given Survey Results, at 95-percent Confidence Level

Reported Result (Percent)	Size Class (Percent)			
	Large ($n = 73$)	Medium ($n = 124$)	Small ($n = 279$)	All ($N = 476$)
1	± 1.7	± 1.6	± 1.2	± 1.0
10	± 5.2	± 4.6	± 3.5	± 2.7
20	± 6.9	± 6.1	± 4.7	± 3.6
30	± 7.9	± 7.1	± 5.4	± 4.1
40	± 8.4	± 7.5	± 5.7	± 4.4
50	± 8.6	± 7.7	± 5.9	± 4.5
60	± 8.4	± 7.5	± 5.7	± 4.4
70	± 7.9	± 7.1	± 5.4	± 4.1
80	± 6.9	± 6.1	± 4.7	± 3.6
90	± 5.2	± 4.6	± 3.5	± 2.7
99	± 1.7	± 1.6	± 1.2	± 1.0

[1] Hubert Blalock, *Social Statistics* (Wiley, 1960), pp. 150–151; Alan Stuart, "Standard Error for Percentages," in *Applied Statistics,* Vol. 12, No. 2, (1962–1963), pp. 87–88.

Appendix B

THE INVESTMENT MANAGEMENT PROCESS

ADDITIONAL DETAIL

The tables in Appendix B elaborate on material reported in Tables 2.1, 2.2, and 2.4, respectively, in the text. These appendix tables report the percentage of foundations that indicated a given actor played an "extremely important" role in the phase of the investment process covered in the table. Since a given foundation could identify more than a single actor as "extremely important" in the particular phase of the process, the percentages noted can exceed 100 percent and are not additive. The figures are useful, however, to show the relative "votes" each of the indicated actors received as "extremely important" participants in the process.

Table B-1
Key Participants in Setting Investment Objectives

Actor	Percent of Foundations Identifying Participant as Playing "Extremely Important" Role[a]		
	All Foundations (*N* = 22,294)	Large Foundations[b] (*n* = 197)	Small Foundations[c] (*n* = 279)
Trustees	46.4	51.3	46.2
Board chairman	24.0	34.9	23.7
Investment committee	16.4	50.0	15.5
Executive director/president	17.1	28.4	16.9
Other foundation staff	7.3	12.3	7.2
Bank or trust department	19.0	17.9	19.0
Other outside managers	9.5	14.7	9.3
Other	2.5	3.6	2.5

Source: Salamon/Voytek Foundation Investment Study.

[a] Figures are not additive; more than one participant can play an "extremely important" role.

[b] Assets of $10 million or more.

[c] Assets of less than $10 million.

Table B-2
Who Sets Foundation Investment Strategy?

Actor	Percent of Foundations Identifying Actor as "Extremely Important"[a]		
	All Foundations (*N* = 22,294)	Large Foundations[b] (*n* = 197)	Small Foundations[b] (*n* = 279)
Trustees	37.7	40.1	37.6
Board chairman	24.8	26.0	24.7
Investment committee	15.5	48.2	14.7
Executive director/president	17.0	23.3	16.9
Other staff	10.1	12.2	10.1
Bank or trust department	21.5	20.9	21.5
Other outside managers	12.7	28.9	12.2

Source: Salamon/Voytek Foundation Investment Study.

[a] Figures are not additive; more than one participant can play an "extremely important" role.

[b] See Table B-1.

Table B-3
Who Handles Day-to-Day Foundation Portfolio Management?

Actor	Percent of Foundations Identifying Actor as "Important" or "Extremely Important"[a]		
	All Foundations (*N* = 22,294)	Large Foundations[b] (*n* = 197)	Small Foundations[b] (*n* = 279)
Trustees	34.8	22.8	35.1
Board chairman	26.4	22.3	26.5
Investment committee	18.4	31.5	17.9
Executive director/president	23.4	26.9	23.3
Other staff	17.3	24.4	17.2
Bank or trust department	27.1	34.0	26.9
Other outside managers	19.0	57.4	17.9

Source: Salamon/Voytek Foundation Investment Study.

[a] Figures are not additive; more than one participant can play an "extremely important" role.

[b] See Table B-1.

Appendix C

SURVEY FORM

UNWEIGHTED SURVEY RESULTS

FOUNDATION FISCAL POLICIES AND PRACTICES PROJECT

CONDUCTED BY THE URBAN INSTITUTE FOR THE COUNCIL ON FOUNDATIONS

October, 1985

Dear Foundation Executive:

The financial well-being of the nation's private foundations is of great concern to those who manage foundations, those who seek funds from them, and the wider public that values their contributions. Unfortunately, however, very little is known about what has been happening to foundation assets, how these assets are managed, or how foundations have responded to recent changes in federal payout requirements. As a result, it is difficult to assess foundation investment performance, to assist foundations in their financial management, or to respond to inquiries from Congress and elsewhere about the effects of public policy changes upon foundations.

To remedy this, an effort is being made to gather a systematic body of information about foundation investment and payout practices. The enclosed form is a crucial part of this effort.

As members of the Advisory Committee overseeing this project, we urge you to complete this form as fully and as quickly as possible and to return it in the enclosed stamped envelope to The Urban Institute, a respected, nonpartisan research organization, which is carrying out this study on behalf of the Council on Foundations. If you cannot complete all of the items, complete the ones that you can. If you have questions or need help, please call either Ken Voytek or Elizabeth Boris at the numbers listed below, or any of us. YOU MAY BE ASSURED THAT ALL RESPONSES WILL BE KEPT STRICTLY CONFIDENTIAL AND THAT ONLY OVERALL TOTALS WILL BE REPORTED.

As we make this request, we know how busy you are and how burdened you are with requests for information. However, we feel it is essential for the foundation community and for the nation as a whole to have the data requested here in order to improve the well-being of foundations and to improve public policy in this field. Thank you in advance for assisting the foundation community in this way.

James A. Joseph, President
Council on Foundations

William L. Bradley, Ret.
Edward Hazen Foundation
Committee Chair

Humphrey Doermann
The Bush Foundation

Jane Dustan
Foundation for Child
 Development

Robert M. Frehse, Jr.
Hearst Foundations

Dorothy A. Johnson
Council of Michigan Foundations

Loren D. Ross
The Russell Sage Foundation

John Simon
Yale Program on Nonprofit
 Organizations

Patricia Carry Stewart
Edna McConnell Clark
 Foundation

Robert E. Swaney
Charles Stewart Mott
 Foundation

William R. Walsh
The Robert Wood
 Johnson Foundation

Louis Winnick
The Ford Foundation

Please refer any questions to:

Kenneth P. Voytek
The Urban Institute
(202) 857-8628

Elizabeth T. Boris
Council on Foundations
(202) 466-6512

Please return the completed survey in the enclosed postage paid envelope to:

Foundation Fiscal Policies and Practices Project
Lester M. Salamon, Director
The Urban Institute
2100 M Street, N.W.
Washington, D.C. 20037

ALL INDIVIDUAL SURVEY RESPONSES WILL BE KEPT STRICTLY CONFIDENTIAL. INFORMATION WILL BE REPORTED ONLY FOR GROUPS OF ORGANIZATIONS.

Please answer all survey questions as completely as possible. If you wish to comment on any of the questions or qualify your answers, use the right margins or the back page of this form. Please call the persons listed on the front cover if you have any questions. If you would like a summary of the survey results, please make sure that your address is correctly listed on the back page of this form.

FOR YOUR INFORMATION . . . . .

THE URBAN INSTITUTE is a nonprofit, research and educational organization established in Washington, D.C. in 1968. Its staff investigates the social and economic problems confronting the nation and public and private efforts to alleviate such problems. Its work is supported by government agencies, corporations, and private foundations.

THE COUNCIL ON FOUNDATIONS, founded in 1949, is a national membership organization of grantmaking institutions, and is dedicated to promoting responsible and effective grantmaking, maintaining a supportive environment for philanthropy, and encouraging collaboration among grantmakers and the formation of new foundations. Its 1,000 members include independent, community, company-sponsored, and private operating foundations, as well as corporate giving programs and trust companies.

BACKGROUND INFORMATION

1. There are four conventional types of foundations. Please identify the one definition that best describes your foundation. [Circle the letter of the single best response.]   **n=478**

    A. INDEPENDENT: fund or endowment designated as a private foundation under tax law whose primary function is the making of grants.   **95.8%**

    B. COMMUNITY: public foundation deriving funds from many donors rather than a single source. Grant programs are generally directed toward the immediate locality or region. The governing board is broadly representative of the community it serves.

    C. OPERATING: fund or endowment certified by the IRS whose primary purpose is to operate programs as opposed to making grants.   **4.2%**

    D. COMPANY-SPONSORED: private foundation organized under tax law for the purpose of making grants and deriving its funds from a profit-making company or corporation.

NOTE: If you circled D above do not complete the survey beyond this point. However, please return this form in the enclosed envelope to The Urban Institute, 2100 M Street, N.W., Washington, D.C. 20037. Thank you for your assistance and cooperation.

---

2. In what time period was your foundation established? [Circle number of your answer.]   **n=470**

1	BEFORE 1900	**1.7%**
2	1900–1919	**3.0%**
3	1920–1939	**14.9%**
4	1940–1949	**15.5%**
5	1950–1959	**21.9%**
6	1960–1969	**22.6%**
7	1970–1979	**16.2%**
8	1980 OR LATER	**4.3%**

3. Which legal device was used to create the foundation? [Circle number of your answer.]   **n=478**

1	CHARITABLE CORPORATION	**59.2%**
2	TRUST	**32.2%**
3	OTHER (specify)	**4.6%** _____
8	DON'T KNOW	**4.0%**

4. Does the foundation have an endowment? [Circle number of your answer.]   **n=478**

1	YES	**90%**
2	NO	**10%**

5a.) Is there a scheduled termination date for your foundation? [Circle number of your answer.]   **n = 478**

    1 YES **1.9%**
    2 NO (If no, go to Q. 6) **98.1%**

  b.) If yes, in what year is this scheduled to occur? [Specify year.]

—— —— —— ——

6.   When does the foundation's fiscal year end? [Specify month and day.]

_____ _____
MONTH   DAY

7.   How many full-time equivalent paid employees did your foundation have as of September 1, 1985? [Circle number of your anwer.] **n = 474**

1	ZERO	**54.4%**
2	1-5 FULL-TIME EQUIVALENT STAFF	**31.2%**
3	6-10 FULL-TIME EQUIVALENT STAFF	**6.5%**
4	11-25 FULL-TIME EQUIVALENT STAFF	**4.0%**
5	26-50 FULL-TIME EQUIVALENT STAFF	**1.9%**
6	51 AND ABOVE FULL-TIME EQUIVALENT STAFF	**1.9%**

[Note: Two half-time staff represent one full-time equivalent.]

SPENDING POLICY

This section seeks information on the process through which spending or payout policies are established and the relationship between payout and investment policies.

8.a.) Please describe briefly how your foundation determines its payout or spending level (for example, is the payout level set at a certain percentage of assets regardless of investment performance or is the expected investment return determined first and the target payout rate derived from it)?

19.8%	Pay out what the law requires
10.7	Pay out slightly more than law requires
28.0	Payout shaped mainly by investment performance
8.9	Payout shaped mainly by program goals
28.0	Pay out all investment income
0.9	Spending determined by trust agreement
2.9	Spending determined by gifts
0.9	No established policy

       n = 450

  b.) Has the procedure for determining your payout rate changed since 1982, when the federal payout requirement was modified? [Circle number of your answer.]

    1 YES **19.1%**
    2 NO (If no, go to Q. 10) **80.9%**
                n = 476

c.) If yes, what was the procedure used to determine payout or spending levels prior to 1982?

24.5%	Pay out what the law requires
6.6	Pay out slightly more than law requires
25.2	Payout shaped mainly by investment performance
8.0	Payout shaped mainly by program goals
30.5	Pay out all investment income
0.9	Spending determined by trust agreement
3.4	Spending determined by gifts
0.9	No established policy

9. If your spending policy has changed since 1982, how important was the change in the payout requirement legislated in the 1981 Economic Recovery Tax Act in leading to this change? [Note: Prior to 1982, the annual payout requirement for private foundations was the greater of (a) adjusted net income or (b) 5 percent of the "fair market" value of assets. In the 1981 Economic Recovery Tax Act, the requirement to pay out adjusted net income if it exceeded 5 percent of asset value was eliminated.] [Circle the number of the single best response.]

1	EXTREMELY IMPORTANT	**42.2%**	**n = 90**
2	FAIRLY IMPORTANT	**27.8%**	
3	NOT TOO IMPORTANT	**16.7%**	
4	NOT IMPORTANT AT ALL	**10.0%**	
8	DON'T KNOW/NOT SURE	**3.3%**	

10. How important were the following factors in affecting your foundation's payout level during the past year? [Circle one number for each factor.]   **n = 478**

	Extremely Important	Fairly Important	Not Too Important	Not Important At All	Not Applicable
A. ECONOMIC CONDITIONS (e.g., inflation)	1 **8.8%**	2 **20.3%**	3 **24.3%**	4 **46.7%**	8
B. ACTUAL INVESTMENT PERFORMANCE	1 **31.6**	2 **28.2**	3 **13.8**	4 **26.4**	8
C. FINANCIAL MARKET CONDITIONS	1 **11.1**	2 **22.6**	3 **25.9**	4 **40.4**	8
D. PAYOUT REQUIREMENT	1 **28.0**	2 **18.4**	3 **15.7**	4 **37.9**	8
E. PROGRAM NEEDS	1 **32.4**	2 **27.2**	3 **9.6**	4 **30.8**	8
F. CUTBACKS IN GOVERNMENT SPENDING/SERVICE LEVELS	1 **4.4**	2 **14.6**	3 **18.4**	4 **62.6**	8
G. OTHER (Specify)	1 **4.2**	2 **1.9**	3 **0.2**	4 **91.2**	8

FOUNDATION INVESTMENT OBJECTIVES

This set of questions deals with the investment objectives of your foundation. Investment objectives refer to the long-term investment goals of the foundation.

11. Which one of the following statements best describes your foundation's principal investment objective? Circle the number of the one answer in the first column that best describes your *current* investment objective? If possible, circle the number in the second column that best describes your investment objective *prior to 1982*. [Circle only one number in each column.] Percentages figures with "DK" *taken out*.

CURRENT (n=468)	PRE-1982 (n=437)	
1 **56.6%**	1 **48.3%**	TOTAL RETURN: Maximizing the sum of appreciation and investment income (interest, dividends, rents, royalties).
2 **18.2%**	2 **20.4%**	INCOME MAXIMIZATION: Maximizing investment income but not necessarily appreciation.
3 **5.8%**	3 **8.7%**	CAPITAL APPRECIATION: Maximizing appreciation but not necessarily investment income.
4 **10.7%**	4 **13.0%**	MAINTAIN REAL VALUE OF ASSETS: Protecting endowment from depreciation (i.e. loss of value).
5 **5.3%**	5 **5.5%**	PROGRAM SPENDING LEVEL: Achieving desired target distribution level.
6 **3.4%**	6 **4.1%**	OTHER (specify) _____
8	8	DON'T KNOW

12. If your foundation's investment objective has changed since 1982, how important was the change in the payout requirement legislated in the 1981 Economic Recovery Tax Act in leading to this change? [Circle the number of the single best response.]   n=115

1 EXTREMELY IMPORTANT   **17.4%**
2 FAIRLY IMPORTANT   **26.1%**
3 NOT TOO IMPORTANT   **18.3%**
4 NOT IMPORTANT AT ALL   **23.5%**
8 DON'T KNOW/NOT SURE   **14.8%**

13. How important a role do the following participants and others you consider relevant play in setting your foundation's investment objectives? [Circle one number for each particpant.]   **n = 478**

	Extremely Important	Fairly Important	Not Too Important	Not Important At All	Not Applicable
A. FOUNDATION TRUSTEES	1 **48.1%**	2 **22.2%**	3 **10.5%**	4 **19.2%**	8
B. BOARD CHAIRMAN	1 **28.0**	2 **15.9**	3 **9.4**	4 **46.7**	8
C. INVESTMENT COMMITTEE	1 **29.3**	2 **9.0**	3 **3.1**	4 **58.6**	8
D. EXEC. DIRECTOR/PRESIDENT	1 **21.8**	2 **14.4**	3 **9.2**	4 **54.6**	8
E. OTHER FOUNDATION STAFF (e.g., Treasurer/Financial Officer)	1 **9.2**	2 **11.3**	3 **10.5**	4 **69.0**	8
F. BANK OR TRUST DEPARTMENT	1 **18.6**	2 **7.1**	3 **4.0**	4 **70.3**	8
G. OTHER OUTSIDE INVESTMENT MANAGER(S)	1 **11.5**	2 **10.9**	3 **4.6**	4 **73.0**	8
H. OTHER (Specify) _____	1 **2.9**	2. **1.9**	3 **0.6**	4 **94.6**	8

14. What rate of return are you seeking to earn over the next three years on your investment assets before adjusting for inflation, taking into account both appreciation and earnings? [Circle number of the single best response.]

1	3-5 PERCENT	**2.6% n = 460**
2	6-9 PERCENT	**42.6%**
3	10-12 PERCENT	**39.6%**
4	13 PERCENT AND ABOVE	**15.2%**

15. Some investments involve greater chance for gain, but also greater risk of loss than others. Over the last 75 years, for instance, the stock market has had an *average* annual rate of return of approximately 9.1%. However, the year-to-year volatility over these 75 years has ranged from a 54.5% increase to a loss of 43.6%. Bonds have generally been less volatile, but still do vary. While no one complains about the ups, the tolerance for downside risks can vary greatly. Generally speaking, what is the tolerance for such *downside risk* in your foundation? [Circle number of the one best response.]   **n = 458**

1 VERY LOW (LESS THAN THE POPULAR BOND MARKET AVERAGES)   **23.1%**
2 MODERATELY LOW (ABOUT EQUAL TO THE POPULAR BOND MARKET AVERAGES)   **40.4%**
3 MODERATELY HIGH (ABOUT EQUAL TO THE POPULAR STOCK MARKET AVERAGES)   **32.3%**
4 VERY HIGH (GREATER THAN THE POPULAR STOCK MARKET AVERAGES)   **4.1%**

16a.) Over what time period does your foundation set its investment objectives and assess its investment performance? Answer by circling the appropriate number in column 1 for the timeframe currently used, and the appropriate number in column 2 for the timeframe used prior to 1982. [Circle only one number in each column.]

(n = 433) CURRENT	(n = 392) PRE-1982	
1 44.3%	1 47.2%	ONE YEAR PERIOD
2 17.3%	2 13.8%	THREE YEAR PERIOD
3 24.2%	3 23.7%	FIVE YEAR PERIOD (i.e., market cycle)
4 4.6%	4 5.1%	TEN YEAR PERIOD
5 9.5%	5 10.2%	OTHER (specify)_____
8	8	DON'T KNOW

b.) If your foundation's time horizon has changed since 1982, how important was the change in the payout requirement legislated in 1981? [Circle number of the single best response.] n = 96

1	EXTREMELY IMPORTANT	13.5%
2	FAIRLY IMPORTANT	10.4%
3	NOT TOO IMPORTANT	21.9%
4	NOT IMPORTANT AT ALL	24.0%
8	DON'T KNOW/NOT SURE	30.2%

INVESTMENT STRATEGY

This set of questions deals with the investment strategies of your foundation and the processes for setting them. Investment strategies translate the broad investment objectives of the foundation into a series of guidelines and policies providing general direction on such matters as asset composition (e.g., relative shares of stocks, bonds, and other investment assets in the portfolio), asset quality, and the choice of either internal or outside investment management.

17. How important a role do the following participants and others you consider relevant play in setting your foundation's investment strategies? [Circle one number for each participant.]

	Extremely Important	Fairly Important	Not Too Important	Not Important At All	Not Applicable
A. FOUNDATION TRUSTEES	1 **38.5%**	2 **24.9%**	3 **14.2%**	4 **22.4%**	8
B. BOARD CHAIRMAN	1 **25.1%**	2 **16.7%**	3 **10.5%**	4 **47.7%**	8
C. INVESTMENT COMMITTEE	1 **27.8%**	2 **9.4%**	3 **3.3%**	4 **59.4%**	8
D. EXEC. DIRECTOR/PRESIDENT	1 **19.7%**	2 **14.9%**	3 **9.8%**	4 **55.6%**	8
E. OTHER FOUNDATION STAFF (e.g., Treasurer/Financial Officer)	1 **10.9%**	2 **10.5%**	3 **8.2%**	4 **70.5%**	8
F. BANK OR TRUST DEPARTMENT	1 **21.5%**	2 **6.1%**	3 **2.9%**	4 **69.5%**	8
G. OTHER OUTSIDE INVESTMENT MANAGER(S)	1 **19.2%**	2 **9.0%**	3 **2.3%**	4 **69.5%**	8
H. OTHER (Specify) _____	1 **3.3%**	2 **2.5%**	3 **0.6%**	4 **93.5%**	8

18. Does your foundation have formal guidelines governing any of the facets of investment portfolio composition listed below? For each possible guideline listed below, please circle the number next to "yes" or "no" to indicate whether or not your foundation has a *formal* guideline or *directive* of this sort. In addition, for those directives which apply, please fill in any blanks and indicate whether you think the guideline has improved the foundation's investment performance, impaired performance, or had no impact on performance.

	Formal Guideline or Directive n = 478	Apply	Impact on Investment Performance?			
			Im-proved	Im-paired	No Effect	Don't Know
A.	No more than _____% of assets in COMMON STOCK	1 Yes **28.9%** 2 No **71.1**	1 **54.3%**	2 **4.3%**	3 **11.6%**	8 **29.7%**
B.	No less than _____% of assets in COMMON STOCK	1 Yes **15.3%** 2 No **84.7**	1 **53.4%**	2 **1.4%**	3 **11.0%**	8 **34.2%**
C.	No more than _____% of assets in FIXED INCOME SECURITIES (e.g., bonds)	1 Yes **22.4%** 2 No **77.6**	1 **54.2%**	2 **0.9%**	3 **10.3%**	8 **34.6%**

Formal Guideline or Directive n = 478	Apply	Impact on Investment Performance?			
		Im-proved	Im-paired	No Effect	Don't Know
D. No less than ____% of assets in FIXED INCOME SECURITIES	1 Yes 16.1% 2 No 83.9	1 51.9%	2 3.9%	3 10.4%	8 33.8%
E. No more than ____% of assets in CASH OR CASH EQUIVALENTS	1 Yes 13.2% 2 No 86.8	1 34.9%	2 3.2%	3 20.6%	8 41.3%
F. No less than ____% of assets in CASH OR CASH EQUIVALENTS	1 Yes 10.0% 2 No 90.0	1 29.2%	2 8.3%	3 16.7%	8 45.8%
G. No more than ____% of assets in COMMODITIES	1 Yes 11.5% 2 No 88.5	1 12.7%	2 0.0%	3 14.5%	8 72.7%
H. No more than ____% of assets in REAL ESTATE	1 Yes 11.1% 2 No 88.9	1 17.0%	2 3.8%	3 13.2%	8 0.0%
I. No more than ____% of assets in FOREIGN SECURITIES	1 Yes 9.4% 2 No 90.6	1 11.1%	2 6.7%	3 6.7%	8 75.5%
J. No more than ____% of assets in PUTS, CALLS, OPTIONS, FUTURES	1 Yes 11.5% 2 No 88.5	1 10.9%	2 3.6%	3 7.3%	8 78.2%
K. No assets in COMPANIES MAKING: _____ (e.g., armaments, nerve gas, cigarettes)	1 Yes 2.7% 2 No 97.3	1 7.7%	2 7.7%	3 46.2%	8 38.5%
L. No assets can be in COMPANIES DOING BUSINESS IN THE FOLLOWING COUNTRY(IES) _____	1 Yes 2.1% 2 No 97.9	1	2	3 60.0%	8 40.0%
M. Other: (e.g. restrictions on the quality of investments or the share of stock in a single company, etc.)	1 Yes 12.8% 2 No 87.2	1 26.2%	2 3.3%	3 19.7%	8 50.8%
N. Other: _____	1 Yes 6.1% 2 No 93.9	1 41.4%	2 0	3 24.1%	8 34.5%

19a.) Has the foundation developed formal procedures for the review of its investment strategies? [Circle number of your answer.]

n = 478

1 YES **50.6%**
2 NO (If no, go to Q 20) **49.4%**

b.) If yes, how frequently are the investment strategies of your foundation formally reviewed? [Circle only one number.]

1 MONTHLY **11.6%**
2 QUARTERLY **39.3%**
3 EVERY 6 MONTHS **20.2%**
4 ONCE A YEAR **24.0%**
5 OTHER (Specify) **4.9%** _____

20. How does your current investment strategy differ from what existed prior to 1982? For each change noted below, please circle the number next to yes or no to indicate whether it applies to your foundation. If the change does apply, also circle the appropriate number to indicate how important the 1981 change in the payout requirement was in causing the change in investment strategy.

Nature of Change Since 1981 n=478	Applies?		Role played by change in Payout Requirement Investment Strategy Change			
			Very Important	Important	Not Very Important	Don't Know
A. More reliance on FIXED INCOME ASSETS now than pre-1982	1 Yes 2 No	22.6% 77.4%	1 26.9%	2 36.1%	3 32.4%	8 4.6%
B. More reliance on STOCKS now than pre-1982	1 Yes 2 No	14.0% 86.0%	1 17.9%	2 31.3%	3 32.8%	8 17.9%
C. More reliance on HIGH YIELD/ LOW GROWTH (Low Risk) stocks now than pre-1982	1 Yes 2 No	8.4% 91.6%	1 30.0%	2 32.5%	3 25.0%	8 12.5%
D. More reliance on COMMODITIES, OPTION, or other high risk investments now than pre-1982	1 Yes 2 No	1.5% 98.5%	1 14.3%	2 14.3%	3 71.4%	8 0.0%
E. More concern about SOCIAL OBJECTIVES now than pre-1982	1 Yes 2 No	5.2% 94.8%	1 4.0%	2 24.0%	3 64.0%	8 8.0%
F. More emphasis on SHORT-TERM INVESTMENTS now than pre-1982	1 Yes 2 No	15.3% 84.7%	1 12.3%	2 38.4%	3 30.1%	8 19.2%
G. Other: _____	1 Yes 2 No	4.0% 96.0%	1 26.3%	2 21.1%	3 47.4%	8 5.3%
H. Other: _____	1 Yes 2 No		1 40.0%	2 20.0%	3 20.0%	8 20.0%

21. How important were the following factors in affecting your foundation's investment strategies over the last year? [Circle one number for each factor.]
    n=478

	Extremely Important	Fairly Important	Not Too Important	Not Important At All	Not Applicable
A. FINANCIAL MARKET CONDITIONS	1 32.0%	2 33.3%	3 8.2%	4 26.6%	8
B. ECONOMIC CONDITIONS	1 22.2%	2 33.7%	3 13.4%	4 30.8%	8
C. CHANGE IN INVESTMENT OBJECTIVES	1 6.1%	2 10.0%	3 18.6%	4 65.2%	8
D. PROGRAM NEEDS OR TARGET SPENDING LEVELS	1 10.7%	2 19.5%	3 19.0%	4 50.9%	8
E. ACTUAL INVESTMENT PERFORMANCE	1 26.4%	2 29.9%	3 13.6%	4 30.1%	8
F. OTHER (Specify) _____	1 2.1%	2 0.6%	3 1.0%	4 96.2%	8

PORTFOLIO MANAGEMENT

This section deals with portfolio management practices. Portfolio management refers to the day-to-day operation of the investment function of your foundation and includes activities such as specific security selection, the timing of purchases and sales, and other day-to-day decisions.

22. How important a role do the following potential participants and others you consider relevant play in the day-to-day management of the foundation's investment portfolio? [Circle one number for each participant.]

n = 478

	Extremely Important	Fairly Important	Not Too Important	Not Important At All	Not Applicable
A. FOUNDATION TRUSTEES	1 17.8%	2 12.1%	3 17.4%	4 52.7%	8
B. BOARD CHAIRMAN	1 14.2%	2 10.5%	3 12.3%	4 63.0%	8
C. INVESTMENT COMMITTEE	1 12.8%	2 10.9%	3 10.9%	4 65.5%	8
D. EXEC. DIRECTOR/PRESIDENT	1 13.8%	2 11.1%	3 12.1%	4 63.0%	8
E. OTHER FOUNDATION STAFF (e.g., Treasurer/Financial Officer)	1 10.0%	2 10.0%	3 7.1%	4 72.8%	8
F. BANK OR TRUST DEPARTMENT	1 25.9%	2 4.2%	3 2.7%	4 67.2%	8
G. OTHER OUTSIDE INVESTMENT MANAGER(S)	1 31.4%	2 2.9%	3 1.7%	4 64.0%	8
H. OTHER (Specify)	1 2.1%	2 1.3%	3 1.5%	4 95.0%	8

23a.) What degree of autonomy and discretion (e.g. able to make independent judgments to buy and sell securities) is enjoyed by the person or group with *primary responsibility* for the day-to-day management of the portfolio? [Circle number of the single best response.] **n = 440**

1 COMPLETE AUTONOMY AND DISCRETION   **50.2%**
2 HIGH DEGREE OF AUTONOMY AND DISCRETION   **33.5%**
3 MODERATE DEGREE OF AUTONOMY AND DISCRETION   **10.6%**
4 VERY LITTLE AUTONOMY AND DISCRETION   **2.6%**
5 NO AUTONOMY AND DISCRETION   **3.1%**

b.) Explain briefly: _____

_____

_____

24. What percentage of the foundation's total investment portfolio is externally managed (including banks or trust departments and investments in mutual funds)? [Specify in percentage terms] **n = 465**

	0% Outside = **43.0%**	
— — — %	1-33% = **3.9%**	67.1% and above = **48.8%**
IF NONE, PROCEED TO Q. 26.	33.1%-67% = **4.3%**	

25. Please provide the information requested in the table below for each outside organization (e.g., investment management firm, mutual fund, bank) that your foundation uses in managing its investment assets. (NOTE: Attach additional page if necessary.) [Fill in the information requested below for each outside organization used.]

TYPE OF ORGANIZATION[1]	% OF ASSETS MANAGED	ASSET TYPE MANAGED[2]	MANAGEMENT STYLE[3]
No. of managers (**n = 461**)			
0 managers = **43.4%**			
1 manager = **39.7%**			
2-3 managers = **9.1%**			
4 managers or more = **7.8%**			

NOTES: 1. For example, Investment Management Firm, Mutual Fund, Bank or Trust Department, Consultant, and the like.
2. Include here asset types such as stocks, bonds, real estate, etc.
3. For example, income-oriented, growth, market-timer, etc.

26a.) Does your foundation currently use outside investment consultant(s) (as distinct from outside managers)? [Circle number of answer.]   **(n = 478)**

1 YES   **24.1%**
2 NO (If no, go to Q. 27)   **75.9%**

b.) If yes, please describe their role and function: _____

_____

_____

_____

_____

27. Which of the following statements best describes the investment management pattern of your foundation?
[Circle number of the single best response.]   **n = 467**

    1   Investment portfolio is *passively managed* with relatively little turnover, review by foundation officials or board members on only a quarterly basis or less, and little use of outside data and analysis on the part of foundation officials.   **34.9%**

    2   Investment portfolio is *managed moderately actively,* with limited turnover, review of performance at most on a monthly or quarterly basis by foundation officials or trustees, and limited use of external data or analysis.   **45.8%**

    3   Investment portfolio is *very actively managed,* with considerable turnover and activity, weekly or monthly review of performance, constant reevaluation of investment strategies in light of market conditions, and heavy use of outside data and analysis.   **19.3%**

28a.) Does your foundation currently use a "master custodial service" (i.e., a service to record transactions, physically handle securities, and/or compute the rates of return)? [Circle number of your answer.]

    1 YES   **36.2%**   n = 478
    2 NO (If no, go to Q. 29)   **63.8%**

  b.) If yes, about how much money was paid for this service in Fiscal Year 1984? [Estimate amount to nearest hundred dollars.]

      $ _____

29. Does the foundation subscribe to any investment performance reporting service? [Circle number of your answer.]

    1 YES   **22.2%**
    2 NO   **77.8%**   n = 478

30. How important were the following factors in affecting the day-to-day management of the investment portfolio over the last year? [Circle one number for each factor.]

n = 478	Extremely Important	Fairly Important	Not Too Important	Not Important At All	Not Applicable
A. FINANCIAL MARKET CONDITIONS	1 **31.4%**	2 **31.4%**	3 **6.9%**	4 **30.3%**	8
B. ECONOMIC CONDITIONS	1 **23.2%**	2 **31.0%**	3 **11.3%**	4 **65.5%**	8
C. CHANGE IN INVESTMENT OBJECTIVES	1 **4.6%**	2 **8.2%**	3 **18.0%**	4 **69.2%**	8
D. CHANGE IN INVESTMENT STRATEGY	1 **8.2%**	2 **10.5%**	3 **17.4%**	4 **64.0%**	8
E. PROGRAM NEEDS OR TARGET SPENDING LEVELS	1 **9.0%**	2 **17.6%**	3 **14.2%**	4 **59.2%**	8
F. ACTUAL INVESTMENT PERFORMANCE	1 **21.8%**	2 **25.3%**	3 **15.3%**	4 **41.8%**	8
G. OTHER (Specify)					
_____	1 **2.3%**	2 **0.4%**	3 **0.6%**	4 **96.7%**	8

31. How important in your foundation's investment decisions are comparisons between your foundation's performance and the performance of other foundations or investment pools (e.g., pension funds, educational endowments)? [Circle number of the single best response.] **n = 471**

    1 EXTREMELY IMPORTANT   **2.5%**
    2 FAIRLY IMPORTANT   **17.6%**
    3 NOT TOO IMPORTANT   **28.0%**
    4 NOT IMPORTANT AT ALL   **45.2%**
    8 DON'T KNOW/NOT SURE   **6.6%**

32a.) Does any portion of your investment portfolio take the form of assets that you do not feel free to manage actively, either because a sale would significantly depress the market value, the assets represent excess business holdings, closely held stock, control stock, or for any other reason? [Circle number of answer.] **n = 478**

    1 YES **16.7%**
    2 NO (If no, go to Q. 33)   **83.3%**

    b.) If yes, what percentage of your investment assets fall into this category? [Specify to nearest whole percentage.]

    — — — %

    c.) How important is the presence of such assets to the current investment performance of your foundation? [Circle number of the single best response.] **n = 85**

    1 EXTREMELY IMPORTANT   **41.2%**
    2 FAIRLY IMPORTANT   **29.4%**
    3 NOT TOO IMPORTANT   **15.3%**
    4 NOT IMPORTANT AT ALL   **10.6%**
    8 DON'T KNOW/NOT SURE   **3.5%**

33a.) Does your foundation currently own any investment assets to which it applies a "blockage" or discount factor in determining the fair market value of the portfolio? [Circle number of your answer.] **n = 478**

    1 YES **6.5%**
    2 NO (If no, go to Q. 34)   **93.5%**

    b.) If yes, what percentage of your total investment portfolio falls into this category? [Specify to nearest whole percentage.]

    — — — %

34. How much money was spent *internally* for the operation of the investment function of your foundation in Fiscal Year 1984? (Include the costs associated with internal portfolio management as well as oversight of outside portfolio management. Include salaries of professional and support personnel and other direct costs, but exclude overhead.) [Estimate amount to nearest hundred dollars.]

    $ _____

35. How much money did your foundation spend on *external management fees* in Fiscal Year 1984? Exclude brokerage fees, other transaction costs, and custodial fees. [Estimate amount to nearest hundred dollars.]

    $ _____

IMPACT OF REGULATIONS

36. How important an effect have the following federal and state regulations governing the operation of foundations had on the investment performance of your foundation over the last two years? [Circle one number for each of the following regulations.]

	Extremely Important	Fairly Important	Not Too Important	Not Important At All	Not Applicable
A. EXCESS BUSINESS HOLDINGS n = 478 [I.R.C. 4943(C)(2)(A)]	1 2.9%	2 3.6%	3 9.4%	4 84.1%	8
B. PAYOUT REQUIREMENT [I.R.C. 4942(d)(1)]	1 16.3%	2 18.6%	3 25.5%	4 43.5%	8
C. PROHIBITION ON SELF-DEALING [I.R.C. 4941(d)(a)(A)]	1 5.9%	2 4.6%	3 8.4%	4 81.2%	8
D. "PRUDENT-MAN" RULE	1 14.6%	2 21.1%	3 15.5%	4 48.7%	8
E. STATE REGULATIONS GENERALLY	1 5.6%	2 7.9%	3 21.8%	4 64.6%	8

RATE OF RETURN ON ENDOWMENT PORTFOLIO

37. Please indicate below the rate of return your foundation achieved on its investment portfolio for each of the past six years. If possible, record the "total rate of return" (including both income and appreciation) on a "time weighted" basis, for the calendar year. However, if your foundation does not keep data in this form, or it is available only for selected years, please provide what you can. [Specify to the nearest tenth of a percent.]

Year	Rate Of Return
1979	— — . — %
1980	— — . — %
1981	— — . — %
1982	— — . — %
1983	— — . — %
1984	— — . — %

38. Indicate whether the rates of return reported above are for: [Circle one number.]

   1 FISCAL YEAR
   2 CALENDAR YEAR

39a.) Indicate whether rates of return reported above are based on: [Circle number of the one response that best applies.]

   1 INCOME YIELD ONLY: i.e. interest, dividends, rents, and royalties but not appreciation or depreciation.

   2 TOTAL RETURN: i.e. income plus depreciation or appreciation of assets.

   b.) If you use total return please indicate whether this is: [Circle one number.]

   1 TIME-WEIGHTED
   2 NOT TIME-WEIGHTED

SELECTED INVESTMENT DATA

40. If possible, record in the table below the following two pieces of data that are needed to complete our analysis of foundation investment patterns: (1) the breakdown of your foundation's investment assets among government obligations, corporate bonds, and stocks for the years 1982, 1983 and 1984; and (2) the total costs your foundation incurred in acquiring investment assets for the years 1979, 1980, 1982, 1983, and 1984.

Information Needed	1979	1980	1982	1983	1984
End of year fair market value of: Government obligations [1]	$	$	$	$	$
Corporate bonds [2]	$	$	$	$	$
Stocks	$	$	$	$	$
Acquisition Costs [3]	$	$	$	$	$

NOTES: 1. Includes Federal, state, and municipal bonds. Treasury Bills with more than one year maturity, and other government agency obligations or securities.
2. Includes convertible debentures.
3. Cost of acquiring investment assets at market value in a year. Do not include commissions or other transaction costs. Include the purchase price of all investment assets, even if the asset was subsequently sold in the course of the year.

41. We would greatly appreciate your providing us with the following information from your 1984 990-PF. (If you prefer, please just enclose a copy of pages 1, 2, 5, and 6 of your completed 1984 990-PF along with this survey form instead of filling out this table.)

Item	Amount	Location on 1984 990-PF
Fair Market Value of Assets	$	(p. 1, upper right hand corner)
Gifts	$	(p. 1, part 1, line 1, col. A)
Investment Income	$	(p. 1, part 1, line 12, col. B)
Operating Expenses	$	(p. 1, part 1, ling 26, col. A minus line 25, col. A)
Grants	$	(p. 1, part 1, line 25, col. D)
Net Investment Income	$	(p. 1, part 1, line 27b)
Adjusted Net Income	$	(p. 1, part 1, line 27c)
Cash	$	(p. 2, part 3, line 1, col. C)
Savings and Temp. Investments	$	(p. 2, part 3, line 2, col. C)
Securities	$	(p. 2, part 3, line 10, col. C)
Investments-Land, Bldg., & Equip.	$	(p. 2, part 3, line 11, col. C)
Investments-Mortgage Loans	$	(p. 2, part 3, line 12, col. C)
Investments - Other	$	(p. 2, part 3, line 13, col. C)
Gross Sales Price	$	(p. 5, part 7, line 1, col. F)
Capital Gain Net Income	$	(p. 5, part 7, line 2)
Net Short-Term Capital Gain	$	(p. 5, part 7, line 3)
Minimum Investment Return	$	(p. 5, part 8, line 5)
Qualifying Distributions	$	(p. 6, part 10, line 4)
PRI's	$	(p. 6, part 10, line 1(B))
Set-Aside	$	(p. 6, part 10, line 3(A)+3(B))
Excess Distribution (A)	$	(p. 6, part 11, line 5, col. A)
Excess Distribution (B)	$	(p. 6, part 11, line 10(E))

OPTIONAL ADDITIONAL DATA

IN ORDER TO BUILD A RELIABLE DATA BASE ON FOUNDATION INVESTMENT PERFORMANCE, WE NEED TWO ADDITIONAL BODIES OF DATA *FROM THOSE FOUNDATIONS THAT HAVE IT AVAILABLE.*

(a)   TIME-WEIGHTED TOTAL RATE OF RETURN *BY CLASSES OF ASSETS FOR 1979–1984; AND*

(b)   COMPOSITION OF TOTAL INVESTMENT EARNINGS, 1979–1984.

IF YOUR FOUNDATION HAS EITHER OR BOTH OF THESE BODIES OF DATA AVAILABLE FOR ANY OR ALL OF THE YEARS BETWEEN 1979 AND 1984, PLEASE RECORD IT ON THE TWO TABLES THAT FOLLOW. IF THESE BODIES OF DATA ARE NOT AVAILABLE, PLEASE GO TO QUESTION 45.

42. **OPTIONAL DATA TABLE A: TIME-WEIGHTED RATES OF RETURN BY ASSET TYPE**

If possible, please record the rate of return you achieved for each type of asset noted below for each year listed. **RECORD TIME-WEIGHTED, TOTAL RATES OF RETURN ONLY.** Record the rate to the nearest 1/10th of 1 percent. If data are not available for each year or for all asset classes, complete as fully as possible.

Type of Asset	Time-weighted Rate of Return					
	Year					
	1979	1980	1981	1982	1983	1984
Cash and cash equivalents [1]	__·__%	__·__%	__·__%	__·__%	__·__%	__·__%
Corporate stocks and other equities	__·__%	__·__%	__·__%	__·__%	__·__%	__·__%
Fixed income securities	__·__%	__·__%	__·__%	__·__%	__·__%	__·__%
Corporate bonds [2]	__·__%	__·__%	__·__%	__·__%	__·__%	__·__%
Government obligations [3]	__·__%	__·__%	__·__%	__·__%	__·__%	__·__%
Real estate [4]	__·__%	__·__%	__·__%	__·__%	__·__%	__·__%
Other (Specify) [5] _____ _____	__·__% __·__%	__·__% __·__%	__·__% __·__%	__·__% __·__%	__·__% __·__%	__·__% __·__%
Overall rate of return	__·__%	__·__%	__·__%	__·__%	__·__%	__·__%

NOTES: 1. Cash, demand deposits, certificates of deposit, money market funds, short-term Treasury Bills (less than one year maturity), etc.
   2. Include convertible debentures.
   3. Include federal, state and municipal bonds and other government agency securities.
   4. Appraisal value of real estate held for investment purposes.
   5. Identify other investments and their market values. Commodity investments, venture capital, and other unique investment assets would be included here.

43. **OPTIONAL DATA TABLE B: COMPOSITION OF TOTAL RETURN ON INVESTMENT**

If possible, please record the total dollar amount of earnings from each source of investment income noted below. If possible, record **GROSS EARNINGS** before deducting brokerage fees and other transaction costs.

Source of Return	Amount of Return					
	1979	1980	1981	1982	1983	1984
Dividends	$	$	$	$	$	$
Interest	$	$	$	$	$	$
Rent	$	$	$	$	$	$
Royalties	$	$	$	$	$	$
Realized appreciation (or loss)	$	$	$	$	$	$
Unrealized appreciation (or loss)	$	$	$	$	$	$
Other: _____ _____	$	$	$	$	$	$
TOTAL EARNINGS	$	$	$	$	$	$

44. Indicate whether reported earnings are: [Circle one number.]

    1 GROSS (i.e., before deducting brokerage fees and other transaction costs.)
    2 NET (i.e., after deducting brokerage fees and other transaction costs.)

45. Would you like to receive a summary of the results of this survey?

    1 YES
    2 NO

46. For purposes of possible follow-up and clarification of information provided, we would appreciate your noting below the name, title, and phone number of the person(s) completing this form.

_____
FOUNDATION NAME

_____
NAME

_____
TITLE

(_____) _____
PHONE

47. We would appreciate your noting below any comments you might have on this survey questionnaire or the issues it covers.

_____

_____

_____

_____

_____

_____

_____

_____

Thank you for your cooperation. Please return the completed survey in the enclosed postage paid envelope to:

Foundation Fiscal Policies and Practices Project
Lester M. Salamon, Director
The Urban Institute
2100 M Street, N.W.
Washington, D.C. 20037